# *THE NEW POLITICS OF POPULATION: CONFLICT AND CONSENSUS IN FAMILY PLANNING*

# *THE NEW POLITICS OF POPULATION: CONFLICT AND CONSENSUS IN FAMILY PLANNING*

**Jason L. Finkle**
**C. Alison McIntosh**
*Editors*

**POPULATION AND DEVELOPMENT REVIEW**
**A Supplement to Volume 20, 1994**

**THE POPULATION COUNCIL**
**New York**

**New York Oxford**
**OXFORD UNIVERSITY PRESS**
**1994**

Oxford University Press
Oxford New York
Athens Auckland Bangkok Bombay
Calcutta Cape Town Dar es Salaam Delhi
Florence Hong Kong Istanbul Karachi
Kuala Lumpur Madras Madrid Melbourne
Mexico City Nairobi Paris Singapore
Taipei Tokyo Toronto
and associated companies in
Berlin Ibadan



Published by Oxford University Press, Inc.,
200 Madison Avenue, New York, New York 10016
and The Population Council, Inc., New York



Library of Congress Cataloging-in-Publication Data

The new politics of population : conflict and consensus in family planning / [edited by] Jason L. Finkle and C. Alison McIntosh.
p. cm.
Outgrowth of a seminar held at the Rockefeller Foundation's conference center in Bellagio, Italy in Feb. 1990.
Includes bibliographical references.
ISBN 0-19-521088-3
1. Birth control—Government policy—Congresses. 2. Population policy—Congresses. I. Finkle, Jason Leonard, 1926– .
II. McIntosh, C. Alison.
HQ766.N52 1994
363.9'6--dc20 94-8051
ISBN: 0-19-521088-3 CIP

1 3 5 7 9 8 6 4 2

Printed in the United States of America
on acid-free paper

# CONTENTS

# ACKNOWLEDGMENTS

THIS VOLUME GROWS OUT of a seminar held at the Rockefeller Foundation's conference center in Bellagio, Italy in February 1990. We are grateful to the Foundation for making the Villa Serbelloni available to us for the seminar and for financial support. In particular, we acknowledge our debt to Dr. Mary Kritz, then of the Rockefeller Foundation, for her interest in and support of the project when it was first mooted. We are also deeply indebted to Daniel Pellegrom and Pathfinder International for additional financial support, interest, and for the intellectual contributions of several members of its staff. Barbara Crane's contribution to the conceptualization of the seminar is also acknowledged with gratitude. In addition, we thank Andrzej Kulczycki and Scott Grosse for comments on the articles in this volume.

As editors, we had initially developed a more rigorous structure for the seminar papers than appears in the final volume. Faced with the diversity of disciplines, geographic locations, and professional experiences of the participants, we later realized that we would do better to allow the authors to speak in their own rich and individual voices. We hope readers of this volume will find that the insights thus presented justify our decision.

Although the Editor and Managing Editor of *Population and Development Review* discourage expressions of thanks for their help, we feel compelled to recognize their patience and critical intelligence in guiding the volume to completion.

Jason L. Finkle
C. Alison McIntosh

# INTRODUCTION

# The New Politics of Population

JASON L. FINKLE
C. ALISON MCINTOSH

FOR MOST OF HUMAN HISTORY, the politics of population has rested on the assumption that population size and growth are essential determinants of national power and economic strength (Strangeland, 1904; Eversley, 1959; Glass, 1940; Overbeek, 1974). To be sure, fear of overpopulation has surfaced from time to time, especially among scholars, but in the past these episodes barely disturbed the even tenor of faith in population size as a defense against aggression.[1] Today, population politics has been transformed as governments everywhere have come to see rapid population growth in third world countries as an obstacle to development and have laid aside the old beliefs. In a major shift of emphasis, the old politics of population has been replaced by the politics of family planning.[2]

While the antecedents of this transformation can be traced back at least as far as Malthus, the change itself has crystallized during the second half of the present century. After two world wars in quick succession, the industrialized nations started to revise their now outmoded belief in the direct relationship between population size and military strength, and recognized the greater importance of economic and technological superiority (Wright, 1955, 1958; Schuman, 1948; Cline, 1975).[3] At the same time, rich nations became alarmed at the unprecedented rise in the rate of population growth in developing countries, as rising standards of living and medical and public health measures, many of them developed during World War II, rapidly lowered mortality. These events gradually led the Western leaders to embrace the idea of population control.

With some notable exceptions such as India and Ceylon (now Sri Lanka), governments of developing countries were slower to abandon the old ideas about population. Uncertain of their place in the world and influenced by Marxist ideology to believe that the notion of overpopulation was an artifact of capitalist imperialism and neocolonialism (Knarr, 1976), newly independent states for a time resisted the efforts of the West to influence them to adopt antinatalist policies. Over time, however, this attitude also changed. First in Asia, and only recently in Latin America and Africa, governments became aware that their efforts to provide housing, schools, and jobs for their citizens were being frustrated by rapid population growth. One by one, they have abandoned their former beliefs and are incorporating family planning into their national policies (Chamie, this volume).

In most of the developing world, the subject of the population policy debate is no longer whether family planning programs should be established and promoted, but how such programs are to be implemented. While at first glance these issues may appear to be less incendiary than international rivalries, recent experience has demonstrated that the control, direction, and objectives of family planning policy and programs can engender at least as much controversy as the argument over official support and sponsorship of population control. Whether or not to provide sterilization or abortion, whether unmarried adolescents should be served, and whether demographic targets or financial incentives should be adopted are policy questions that frequently engender intense political debate. No less controversial are the decisions concerning what agencies are to be involved and who is to lead, staff, and fund them. All these questions have generated fierce debate that has spilled over from the scientific and bureaucratic domains into the political arena.

What is it about family planning programs that makes them such targets of disputation? The intimate connections between family planning and sex, reproduction, and the family have always made attempts to influence fertility behavior a sensitive—if not volatile—issue. The controversies over family planning are especially acute because almost all programs are run by governments, or receive government approval and support, and government actions and motives seldom enjoy the confidence of the community. This is a problem particularly in societies with a legacy of arbitrary and exploitative rule, whether under colonial or indigenous rulers. Additionally, public skepticism is reinforced by the low quality of services provided by many governments.

As traditional moral and religious objections to birth control have become more muted, the major thrust of criticism of family planning programs is increasingly along ethical lines. Feminists and other critics have objected to programs that are intended to bring about demographic change. They argue that the demographic rationale encourages programs to rely disproportionately on irreversible and long-term methods that restrict women's control and on hormonal preparations that impose excessive health risks on women. Some critics contend that societal benefits are irrelevant if the needs of individual women are not addressed (Hartmann, 1987; Dixon-Mueller, 1987; Barroso, 1990; Germain, 1987).

While the old demographic rivalries among nations seem to have subsided for the time being, they still manifest themselves at the subnational level among ethnic, religious, and communal groups. Examples abound of cases where such rivalries, or the fear of unleashing them, has impinged on some aspect of population policy formulation or implementation in India or Africa (Miller, 1971; Mazrui, 1971; Kokoli, this volume; Pai Panandiker and Umashankar, this volume). In Nigeria, distortion of ethnic and regional numbers in successive censuses is related to competition for control over the institutions and resources of central government (Ekanem, 1972; Kirk-Greene, 1971; Adepoju, 1981). Subnational rivalry is also evident in Malaysia, where the Malay population is being urged to procreate to assure their continued numerical dominance over the Chinese (Ness, 1993), and in Lebanon where demographic and ethnic rivalries engulfed the country in protracted civil war (Chamie, 1981; Gilmour, 1983). Yet while communal or religious conflict may dissipate political support for family planning, and civil unrest may disrupt the delivery of services, there is less evidence that either factor will lessen the desire for family planning once a demand has been created (see for example Faour, 1989, on Lebanon).

## The changing character of family planning

As Paul Demeny has often reminded us, "selling" contraception is a very different proposition from selling the daily loaf; frequently the demand for family planning has to be created. Over the years, family planning program managers have learned that fertility reduction calls for much more than the distribution of one or two contraceptive methods to women who seek them out in city clinics. To make an impact, and to gain the trust of the community, program managers have had to learn how to motivate women to become acceptors, to educate and counsel them, to provide backup services and alternative methods, to encourage continued use, to offer emergency health care for sick children and referrals for infertility problems.

Indeed, established family planning programs today have become sophisticated and complex operations, often extending to the farthest reaches of the society, with the capacity to enter every home and local market. Family planning programs provide employment for doctors and nurses; midwives; trained birth attendants; community volunteers; communications, logistics, and management specialists; social and biomedical scientists; survey researchers and evaluation specialists. Programs encompass both integrated and categorical family planning services in government and private clinics and hospitals, outreach programs, community-based distribution, social marketing programs, and programs of information and service delivery to women and men in factories and on plantations.

The abundance and variety of services have undoubtedly improved the quality and reach of family planning programs in many countries. What has often escaped observation is that the structural complexity and diversity of activities have not only given programs greater political visibility, but have also made them more vulnerable to the attentions of interest groups and individuals who are intent on shaping programs or parts of them. Substantive decisions about whom to serve or which contraceptives to offer are likely to be scrutinized not only by scientific experts, but also by organized interest groups that are promoting specific agendas. In short, the enhanced significance of family planning in monetary and programmatic terms has had the effect of converting a myriad of technical questions into political ones.

## The emergence of international concern

The decision whether or not to introduce programs in population and family planning became an issue of international concern during the 1960s when a number of forces converged to change the attitudes of political actors in both industrialized and nonindustrialized countries. The first intimations of a change in attitudes toward rapid population growth were associated with disappointment in the international community at the relatively slow rate of progress of development in the third world compared to optimistic expectations. Although many countries had been able to achieve impressive rates of economic growth during the 1950s, these were largely offset by high rates of population growth. As early as 1959, the Food and Agriculture Organization (FAO) drew attention to the growing gap between rich and poor countries in the production and consumption of food, and pointed out that in some countries population growth was outstripping food production (FAO, 1959, 1960; see also Symonds and Carder, 1973). The first UN Development Decade, launched in 1961, also brought disappointment, because external assistance increased more slowly than had been expected. In fact, despite increased private overseas investment, "the net transfer of resources from rich to poor countries virtually dried up" (Symonds and Carder, 1973).

A second factor helping to legitimize the notion of international population assistance was the growing knowledge of demographic trends that became available after the wave of population censuses taken in 1960–61. The census results indicated that populations throughout the world were growing at a much more rapid rate than either the countries themselves or the United Nations Population Division had realized (Gille, 1961; Sauvy, 1963). Moreover, a number of scholarly analyses had appeared that provided an intellectual link between the rapid growth of population and the disappointing rate of development. Prominent among these was the seminal study by Coale and Hoover (1958) which suggested that high fertility acted as a brake on development through its effects on age structure and dependency. Also influential were the United Nations volume *The Determinants and Consequences of Population Trends*, the first edition of which appeared in 1953, and Nelson's elaboration of the theory of the low-level equilibrium trap (1956).[4]

A third factor fueling international concern was the articulation within some poor countries of the belief that population growth was a major obstacle to development. In a number of Asian countries, most notably those of the Indian subcontinent, scholars and political elites had expressed an interest in lowering fertility during the period between the two world wars (Myrdal, 1968, vol. 2; Symonds and Carder, 1973). In these countries, as well as in parts of British-ruled Africa, Egypt, Singapore, and Hong Kong, family planning associations were formed before or soon after World War II. Although India made a modest beginning under the First Five Year Plan, and Egypt started to offer official family planning services in a small way soon after the revolution of 1952, the first population policies were not adopted until the 1960s.[5]

A fourth factor that encouraged both developed and developing countries to view family planning as a feasible proposition was the appearance of a new contraceptive, the intrauterine device (IUD). Unlike other contraceptive methods available at the time, the IUD was thought to be safe, effective, reversible, inexpensive, and easy to administer (Balfour, 1962). Above all, the IUD did not require daily administration nor did it have to be used at the time of coitus. In places where the distribution system and the motivation of users were both weak, the IUD gave promise that effective program implementation might be possible. The IUD played a particularly important role in inducing the governments of India and Pakistan in the mid-1960s to adopt ambitious targets for reducing fertility rates (Finkle, 1972).

While the convergence of these different events, ideas, and perceptions set the stage for the legitimization of official international population assistance, the last and most critical factor was the reversal in 1965 of the United States position on population assistance. During the 1950s, the United States was the dominant world power because of its military, trade, and financial strength, the economic weakness of the West European countries, and the incomplete consolidation of the Communist bloc (Cox and Jacobson, 1973). At this early date, moreover, developing countries still lacked the ability to formulate a united position on development issues—even if they had been inclined to pressure the United Nations or the United States on population control. In short, prior to the mid-1960s, there was no combination of nations willing and able to assume leadership on population or to persuade the United States to assert its influence on behalf of birth control in developing countries.

Much of the explanation for the hesitance displayed by the United States government over the population issue can be traced to American conservatism on sexual issues, the controversies that had surrounded family planning in the United States, and the inflexibility of the Catholic Church on the question of birth control (Piotrow, 1973). By the early 1960s, however, cracks were appearing in the Catholic position. Not only was it apparent that Catholic women were increasingly using artificial methods of fertility regulation (Westoff and Bumpass, 1973; Ryder and Westoff, 1977; Mosher and Goldscheider 1984), but Rome itself appeared to be reconsidering its absolute ban on birth control (Keely, this volume). The arrival on the scene of President John F. Kennedy, whose work in the United States Senate had alerted him to the vitiating effect of population growth on development, provided the necessary impetus for the United States to assume a leadership role. Kennedy's sensitivity to this issue and his appointment of sympathetic advisers in the State Department and the White House paved the way for the United Nations' acceptance in the fall of 1962 of a Swedish draft resolution on population and development that had been held up in the General Assembly for more than a year (Schlesinger, 1965). While both the resolution and the United States position still fell short of endorsing United Nations or United States involvement in technical assistance in family planning, several commentators have judged the Kennedy administration to be the turning point in US attitudes toward population (Schlesinger, 1965; Symonds and Carder, 1973).[6]

In the remaining sections of this essay we focus on the three most significant arenas of political action relative to family planning. First, we examine the role of politics within the United Nations and its specialized agencies. Next, we discuss the role of bureaucratic politics, primarily in national governments. We conclude by drawing attention to the increasing significance of nongovernmental transnational actors in the politics of policy formulation and implementation.

## The United Nations and family planning

The United States was not alone in its reluctance to become directly involved in international population assistance; the United Nations system was also slow to envisage a role for itself in this field. The regulation of population growth was universally regarded as a sensitive political issue, and within the UN Population Commission any thought of intervention was strongly resisted during the 1950s and early 1960s. Some of the most sustained opposition came from Catholic countries and from the Soviet Union, the latter regarding rapid population growth as a consequence of the economic policies of colonial powers. Individual initiatives within the United Nations Educational, Scientific, and Cultural Organization (UNESCO), Food and Agriculture Organization (FAO), International Labor Organization (ILO), and the World Health Organization (WHO) had all failed during the 1950s (Symonds and Carder, 1973) and had left these agencies wary of trying again. WHO, an obvious contender for a leadership position in family planning, had been badly shaken in 1950 when a proposal merely to create an expert committee on "the health aspects of population dynamics" alarmed the Vatican and caused the representatives of several Catholic countries to threaten to withdraw their membership (Finkle and Crane, 1976). Once the United States reversed its position, however, it was able to use its influence and money to involve the United Nations.

The specific mechanism designed to encourage the provision of family planning assistance in the United Nations system was the creation by the secretary-general in July 1967 of a small Trust Fund for Population Activities to which interested donors could contribute. Two years later the fund, renamed the United Nations Fund for Population Activities (UNFPA), was moved into the United Nations Development Programme (UNDP), which already had a worldwide network of in-country representatives who could facilitate requests from member states for technical assistance (Symonds and Carder, 1973; Piotrow, 1973). The fund's newly appointed director, Rafael Salas, was authorized to finance population activities primarily within the specialized agencies. Initially, the fund was not intended to be an executing agency but simply a funding body. It was also expected to be able to act as the coordinator of activities that were to be carried out by the specialized agencies, particularly WHO, ILO, UNESCO, FAO, and the United Nations International Children's Emergency Fund (UNICEF). The idea was that the availability of special funds earmarked for population activities would act as an incentive for the specialized agencies to encourage and assist developing nations to limit their population growth. Additionally, small countries that might not have the desire or capacity to develop their own bilateral programs would be able to make contributions to the fund.

Once committed to the United Nations' population program, the United States rapidly became not only the major source of funding, but also of intellectual stimulation. The special contributions of the United States reflected both its sense of global partnership and its strong commitment to promoting fertility reduction through all available channels. In part, United States donations were also prompted by an unanticipated increase, in late 1967, in congressional appropriations for US population assistance that could not easily be absorbed within the country's bilateral programs and that presented a potential source of political embarrassment in the face of cutbacks in other development programs (Symonds and Carder, 1973). The administration also hoped that channeling funds through the United Nations system would shield the United States from the charges of imperialism, racism, and genocide that were being leveled against Western-inspired population control.

To understand the politics of family planning within the UN system, it is important to realize that the United Nations is not a hierarchical organization, but a loosely connected system of autonomous and quasi-autonomous councils, commissions, and agencies. While the title, Secretary-General of the United Nations, suggests that the incumbent of that post exerts executive control over the system, this is far from the case. Each of the specialized agencies is an independent organization structured similarly to the United Nations itself, with a secretariat, and an executive head whose authority over the agency roughly parallels that of the secretary-general over the central organs in New York. Like the United Nations itself, the specialized agencies are loosely governed by their member states, which meet annually or biennially in a general assembly, and more intensively by a smaller governing body or council elected by the assembly.

Many of the difficulties that beset family planning, no less than other substantive development programs, can be traced to this lack of central control.[7] First, government by the large membership on the principle of one country–one vote means that the secretariats have often been unable to get their programs approved by their general assemblies. Second, despite an almost obsessive concern over coordination, the specialized agencies have jealously guarded their autonomy. As a result, the development system as a whole has been rife with territorialism, competition, and overlapping mandates. While for many years the specialized agencies had considered population as too sensitive an issue to grapple with, once the tide turned, none of them was willing to bow to the direction of UNFPA or any other single agency. Indeed, WHO, which had earlier argued that population was not a health problem but a social and economic one, reversed itself and claimed family planning as primarily a health issue (Finkle and Crane, 1976).

The creation of the Trust Fund for Population was consistent with the approach worked out by the major donors in industrialized countries to gain more control over programs that they considered particularly important. By providing special funds earmarked for specific purposes, which were separate from the regular budgets of the specialized agencies, the donors hoped to avoid having their favorite programs derailed in the general assemblies. One of the first uses of this mechanism was the creation in 1949 of the Expanded Programme of Technical Assistance, later to become UNDP, which was intended as a way of nudging the specialized agencies into doing more in the technical assistance field. In population, the funds provided for WHO's Expanded Programme in Human Reproduction was a way of getting WHO more fully involved in family planning.

Under the vigorous leadership of Rafael Salas, UNFPA was eminently successful in attracting funds and, more importantly, in institutionalizing family planning in a difficult organizational environment. Even Salas, however, was never able fully to overcome all the difficulties of working in the decentralized United Nations system. Although the specialized agencies finally dealt with population and family planning, and were happy to receive UNFPA funds for this purpose, their commitment continued tepid. As might have been expected, their primary loyalties remained true to their core missions. Moreover, even had they wished to, the specialized agencies could not ignore the demands of their constituencies in member states.

These, at least through the 1970s, reflected their preference for assistance in areas they considered more fundamental—primary health care and maternal and child health at WHO and UNICEF, basic education at UNESCO. In spite of the funds provided by UNFPA, the specialized agencies were slow to commit their own resources to population activities. Disappointed with the response of the specialized agencies, and eager to make more efficient use of the resources at his disposal, Salas diverted increasing sums to the direct support of population work by governments and private organizations in developing countries (Ness, 1979).

Like the specialized agencies, UNFPA has been required to maintain a delicate balance between the wishes of its donors and recipients as represented on its governing council. Salas himself made no secret of his interpretation of the fund's purpose as supporting a broader range of activities than the family planning programs preferred by the major donors. Salas constantly reminded his staff of the sovereignty of states and urged them to consider seriously all viable requests from governments once they had developed their own population policies and priorities (Salas, 1976). Responding to many such requests, UNFPA has allocated substantial sums to such "beyond family planning" projects as migration and urbanization, education in population and family welfare, basic data collection, women's status, aging, and research on the determinants of fertility (Salas, 1976). In turn, the breadth of UNFPA's involvement in population and development has prompted its richer donors periodically to remind UNFPA that its first priority should be family planning. In practice, however, UNFPA has consistently allocated approximately 50 percent of its support to family planning, broadly defined.

From UNFPA's inception until 1984, the United States government was by far the largest single donor to the fund. As might be expected, with American money came attempts by the United States Agency for International Development (USAID) to influence UNFPA's policies and direction. This was not altogether an unhealthy development, since it produced a creative tension that helped UNFPA to demonstrate a level of dynamism not generally associated with United Nations organizations. Despite the abrupt cessation in 1985 of United States financial support for UNFPA, both the fund and the United States have recognized the need for continuing dialogue. In the expectation of both sides that the United States will once again become a major contributor to UNFPA, the United States has retained more influence than would otherwise have been expected.

## Bureaucratic politics and family planning programs

As mentioned at the beginning of this essay, a major change has taken place in the arena in which the politics of population has been played out in the last third of the present century. Previously, issues of population size and growth generated debate among members of the ruling elite, the social and political class that bears responsibility for assuring internal stability and well-being and external power and influence. In recent years, the focus of debate has shifted and the political give and take is now primarily centered on controversies over family planning programs. Where formerly the population debate touched primarily on themes of consequence for a nation's rulers, today increasingly it addresses questions of concern to bureaucrats and technical specialists as well as the population that is affected by family planning programs.

A major change in the debate over controlling population growth was precipitated by the appearance of new technologies of contraception in the 1960s. The development of a variety of safe and effective contraceptives that are relatively easy to use and administer has given governments, for the first time in history, the possibility of influencing fertility trends. Today, more and more governments identify their "population problem" as rapid population growth and see family planning as a major means of dealing with the problem (Chamie, this volume). To be sure, the old geopolitical idea that a large and growing population is a sign of national health still survives here and there. However, leaders of developing countries increasingly realize that the well-being of their societies depends upon the ability to provide jobs, schools, housing, and health care for their citizens, all of which tasks are made more difficult by rapid population growth.

The appearance of family planning on the agendas of governments in the third world has brought with it a substantial infusion of monies for the operation of programs. Writing elsewhere (Finkle and McIntosh, 1980; McIntosh, 1983), we have made the point that a population policy is not merely an expression of sentiment, although just such an expression, in the form of a statement of demographic goals, may be the starting point. To be taken seriously, a population policy must also include the elaboration of a course of action by means of which the objectives may be achieved. An agency must be designated or created and endowed with authority to implement the course of action and, finally, an adequate budget must be appropriated to enable the agency to carry out its mandate. The absence of any one of these elements, especially the budget, suggests that the government is not yet fully committed to the policy.

The elaboration of a full-fledged policy—with attainable goals, the expectation of additional resources, and the promise of increased visibility and influence in governmental circles—is almost invariably accompanied by fierce interagency competition for the new resources and the power and resilience that they bring. Competition for the leading role is especially likely in such a field as population in which relevant activities cut across a number of traditionally defined economic and social sectors. There is a strong probability that the new enterprise will find itself at the center of a bureaucratic struggle in which established agencies with more clearly defined turf, specialized professionals, and recognized spheres of action view the newcomer as an interloper and attempt to seize the population domain, or significant parts of it, for themselves. If the new program has been endowed with a committed and energetic leader, well-versed in the bureaucratic politics of government, and accorded strong political support from the top, it may succeed in establishing itself as a autonomous entity; if support and leadership are lacking it may well be engulfed by more-established agencies.[8]

These observations are not merely theoretical or speculative, but are grounded in the experience of a number of countries. In Indonesia, powerful political support and skilled leadership enabled a strong, flexible program to emerge, capable of coordinating the efforts of other government agencies and donors (Warwick, 1986). More often, especially where political support from the top leadership is weak, the bureaucratic infighting is likely to result in fragmentation and duplication of services. Even more debilitating is the uncooperative climate that tends to emerge in these situations. Egypt during the 1960s and 1970s provides an example.

It has been observed by a number of authors that neither President Nasser nor President Sadat was strongly committed to family planning as a solution to Egypt's population growth problems; both preferred to hope that the desert could be made productive and the population dispersed.[9] When government family planning services were initiated in 1965, therefore, they were integrated into the Ministry of Health where they were forced to compete, with only moderate success, for funds, personnel, and other resources. Furthermore, the coordinating body, the Supreme Council for Family Planning, and its secretariat, the Family Planning Board, were more interested in development than in family planning, and regarded their responsibilities as incorporating broader aspects of population policy. The office responsible for family planning within the Ministry of Health remained weak and was unable to assert its leadership in the 1970s as the Supreme Council moved toward establishing an integrated program of population and development. This period witnessed a bureaucratic struggle in which alliances were formed among a number of donor agencies and the heads of various divisions and directorates in the Ministry of Health, each of whom seized the opportunity to enrich the resources available to his/her unit, irrespective of whether it had any formal responsibility for family planning. The family planning office emerged even weaker than before, with a minuscule budget dwarfed by those of its competitors within the agency, and incapable of addressing itself to the numerous problems of overlapping responsibilities, fragmented programs, and chaotic budgetary arrangements (Finkle, 1982).

The situation in the Philippines immediately prior to the announcement of a population policy in 1970 was analogous. In this instance, a coalition of indigenous organizations and foreign donors undertook a multi-pronged campaign to raise awareness of the population growth problem and encourage the government to take action (Warwick, 1982). As part of this coalition, USAID attempted to create a domestic constituency for family planning by funding as many indigenous family planning agencies as it could reach and, in addition, sponsored research and supplied contraceptives. While this was a successful program prior to 1970, greatly raising the visibility of family planning, it became much less functional after the policy was announced and the indigenous agencies entered into competition with each other for additional resources. The competitive environment that developed made it difficult for a strong national program to emerge (Warwick, 1982).[10]

The competitive behavior of organizations in situations where significant new monies become available should not be seen as a form of bureaucratic pathology or as aberrant behavior. Organization theorists long ago pointed out that the primary goal of any organization is to survive and that all organizations devote some proportion of their resources to survival strategies. If an agency is able to secure additional funds, personnel, and matériel, even at the cost of taking on broader responsibilities, it will be in a stronger position to endure if times turn bad.[11] In the population field, the stakes may be raised and the number of stakeholders increased by the presence of foreign donors, each struggling to broaden its sphere of influence.

While organizations frequently pursue new resources in order to enhance their ability to survive, survival should not be understood only as part of a quest for interagency power and influence. To many members of a bureaucracy, new resources mean an increased capacity to advance objectives they consider to be significant and essential goals of social action. In other words, far from being automatons, bureaucracies are suffused with values, goals, and preferred modes of action that they will struggle to promote. In large measure, these characteristics are a product of the organization's most central mandate and the training and socialization of its professionals. Over time, each organization develops its own bureaucratic culture and organizational mission, the protection of which tends to become an important organizational goal in itself. An illustration of this sort of bureaucratic behavior that has had positive results is provided by the Office of Population within USAID, which, under both the Reagan and Bush regimes, functioned in an unsupportive environment yet was able to maintain a strong sense of its mission to increase contraceptive prevalence.

Ministries of health show a particularly strong propensity to protect their organizational boundaries and traditional missions. Despite the evident medical dimension of family planning, the primary socialization of the physicians who staff ministries of health encourages them to see family planning as peripheral. Where governments have tried to locate family planning within health ministries, there have been frequent complaints that family planning diverts resources from such primary professional goals as the reduction of infant mortality, the control of infectious and parasitic diseases, or the extension of services to underserved areas.[12] Within the family planning service itself, professional medical ethics frequently prompts unease and generates conflict over such issues as whether or not sterilization and abortion should be permitted, contraceptives should be supplied to unmarried girls, physical examinations should be performed on new acceptors, or whether it is right to use commercial channels for the distribution of contraceptives. The marginal status of family planning in many ministries of health also puts it at a disadvantage in the competition for attention and resources.[13] More commonly, however, as is the case in many sub-Saharan African programs today, the weakness of family planning programs in health ministries reflects a low level of organizational effectiveness and political commitment to family planning that may be based, in part, on the perceived lack of demand by the public (Caldwell and Caldwell, 1987).

## The politics of family planning program implementation

Family planning specialists have not infrequently sought to portray the implementation of programs as a purely technical and logistical operation. While this position may be adopted in an attempt to shield programs from political controversy, it commonly proves inadequate to the task. The reason for this inadequacy is that program implementation involves not only technical decisions, but also allocative and ethical decisions that tend to provoke political and organizational differences among administrators with differing areas of responsibility within the program. So pervasive are these differences that students of organization have recognized the question of how to ensure that the program designed at the center is implemented as intended in the field as one of the central problems of organizational effectiveness (Selznik, 1949; Smith, 1967, 1985; Wildavsky and Pressman, 1979).

Disparities between the expectations of officials at the top, middle, and local levels are an important source of tension in many programs. As Warwick (1982) has pointed out, the cause of these differences tends to be more structural than personal and it often has to do with the different circles within which officials move. At the highest levels, program leaders tend to be more highly educated, cosmopolitan, and technocratic in orientation than are lower-level officials. Frequently, top-level leaders have received donor-supported foreign training and have imbibed the values of the global population establishment. They are more likely to see rapid population growth as an urgent national problem and to stress the demographic results expected of the program. Many such leaders lack sensitivity to the problems of workers at the local level who encounter the doubts and fears of the target populations. Not well educated or trained, local family planning workers often have to deal with resistance from the population they serve. Frequently, this resistance is reinforced by the preaching of local religious officials, or the advice of local politicians and opinion leaders, whose horizons may be limited (see e.g. Gadalla, 1979; Pai Panandiker and Umashankar, this volume).[14] Officials at intermediate levels are caught in the classical organizational bind of having to face both ways simultaneously (see, for example, Greenhalgh, 1993).

The literature on family planning is replete with examples of controversies whose origin lies in differing perceptions of officials at different levels of the program. One of the clearest comes from Egypt's Population and Development Program. While this program was in effect, family planning was only one of 13 different programs that were supposed to be implemented by village-level workers—an unrealistic burden. The workers, faced with a recalcitrant population who were influenced mainly by conservative sheikhs and mullahs, found it easier to concentrate on programs that were more popular than family planning (Gadalla, 1979). The basic problem was reinforced by the inability of a weak ministry to provide training in family planning to a large enough number of clinic-level physicians who might have acted as a counterweight to local religious leaders (Finkle, 1982). Meanwhile, the few, poorly paid physicians who were in place were more interested in supplementing their inadequate salaries by seeing family planning clients at their private clinics after hours. Had there been a larger cadre of trained and committed family planning professionals at the governorate and district levels, the program might have been more sensitive to the difficulties experienced by the village-level workers.

Intraagency conflict surfaced in the Philippines (Warwick,1982) where sterilization, promoted by the leadership as a low-cost method free of side effects, was resisted by regional and local administrators who felt that they were already overburdened with the problems of implementing existing programs. Regional and local implementers were also distressed about introducing community-based distribution (CBD), as they had serious reservations about the possibility that young couples who had not yet had a child would be able to obtain contraceptives (Warwick, 1982). While CBD programs and sterilization are more widely accepted today, there are reports that the introduction of new methods such as NORPLANT® is just as likely to engender tensions among officials at different levels of the program. A WHO task force researching the introduction of new contraceptive methods has found that service providers who feel insufficiently trained in a new method tend to suggest a method with which they are more familiar. In one country where NORPLANT® has recently been introduced, program officials have voiced concern that greater attention be paid to problems of service delivery and quality of care—especially access to removal of the implant—while officials at higher levels are more interested in the demographic impact of the method (Spicehandler, personal communication, 1992).

Finally, the Maternal and Child Health–Family Planning (MCH-FP) Extension Project in Bangladesh vividly illustrates how insensitivity on the part of central ministries and foreign donors can create difficulties for programs on the ground. In the early 1980s, the Planning Commission of Bangladesh consulted with foreign donors to obtain funding for an experiment in which innovations in service delivery that had been successful in Matlab *thana* were to be introduced into the Ministry of Health program (Phillips, Simmons, and Koblinsky, 1985). The project was to be carried out by ICDDR,B, the international agency that had developed the new strategies. The Ministry of Health felt itself to have been bypassed in the negotiations and for some time remained lukewarm in its support of the project. There are many lessons that can be learned from this case. One, however, stands out: All organizations develop territorialities, and even powerful agencies like the Planning Commission of Bangladesh must take into account the sensitivities of other ministries.

## Family planning in federal systems

At first glance, it might appear that relations between the center and the constituent units in a federal system would be static and without interest in that the divisions of powers between them are constitutionally determined and require special constitutional procedures to amend (Smith, 1985). In reality, center–state relations in federations are as subject to negotiation, bargaining, and change as they are in any other form of government. Indeed, one scholar has suggested that the constitutional incorporation of the regions into the center's decisionmaking procedure is the single feature that distinguishes federated from nonfederated states (King, 1982).

The factors that underlie both the decision to federate and the changing relationships within federations are numerous and diverse. They include historical and ideological forces as well as political and economic preferences, all of which are subject to reinterpretation as conditions change. Whether the push to federate came from the center in an effort to contain the divisive nationalism of formerly autonomous units, or from peripheral units seeking more unified trading relationships or protection from a predatory neighbor, is also important in determining the character of the center–state relationship. Moreover, control of a specific policy domain may not be fully determined by the formal allocation of powers between the center and the peripheral units; central governments in federations, no less than in unitary systems, can find ways to encourage states to adopt or strengthen specific policies, notably by providing additional funds for this purpose.

Like other areas of social policy, population policy and family planning services have felt the impact of center–state relations in a number of federal states. Largely because of the decrease in federal funding for family planning during the Reagan and Bush administrations, the provision of such services for women of low income in the United States is now much more dependent on the level of political commitment to family planning in the individual states (Gold and Guardado, 1988). The division of responsibilities between the federal and Land governments in the former West Germany severely constrained the ability of the federal government to formulate and implement a pronatalist policy during the 1970s and 1980s (McIntosh, 1983). Different levels of commitment to family planning are also evident among the states in Mexico (Cabrera, this volume). The most interesting and difficult problems of center–state relations, however, are those that have dogged India's family planning program since its inception.

The rapid growth of India's population and its anticipated effect on the nation's ability to reduce its poverty were widely discussed in academic and political circles in India before World War II (Myrdal, 1968). Even before independence, the National Planning Committee of the Congress Party under the chairmanship of Jawaharlal Nehru had argued for a population policy, a call that was repeated in the First Five Year Plan of the newly independent government in 1951 (Myrdal, 1968). When the time came to act, however, Nehru's government decided to locate the family planning program within the ministry of health. While the constitution at that time placed health within the jurisdiction of the states, the central government assumed responsibility for funding the family planning program. Using its financial powers, the central government played a vital role in formulating national family planning policy, setting acceptor targets, determining the basic strategies to be adopted, and allocating resources to the different parts of the program.

The decision to integrate family planning with health, and thereby to remove it from the direct control of the central authorities, implanted the fledgling program in a doubly inhospitable environment. First, in India as elsewhere, family planning tended to be regarded as a troublesome interloper in ministries of health, introducing services that few outside of the highest levels considered important, and consuming resources and energies that might have been directed to traditional health programs. Second, senior ministry of health officials in the states, whether generalist administrators of the elite Indian Administrative Service, the Provincial Civil Service, or medical directors, were subject to the oversight of state and local politicians who tended to see few votes in family planning. While the level of public demand for family planning varied from state to state, it was usually weak, and especially so in the large, impoverished, and less advanced states of the north. The combination of bureaucratic and political disinterest combined to create a situation in which family planning came to be perceived as a second-class program, unlikely to launch its officers on distinguished careers and unable to recruit potential high-flyers.[15]

Disappointed with the lack of results, Mrs. Ghandi's government in 1976 changed the constitutional status of family planning, placing it on the Concurrent List where it falls under the joint jurisdiction of the center and the states (Pai Panandiker and Umashankar, this volume). This change means that in case of differences between the center and the states, the center's legislation takes precedence; in practice, however, the center still lacks the resources and personnel to work closely enough with the states to effect change.[16] This is not to say that there are not many local examples of well-run programs in India and that overall progress has not been made in the more affluent states. Despite numerous reorganizations, however, the family planning program remains weak and has yet to show results in the populous north (Dyson and Moore, 1983).[17]

The political and administrative problems of center–state relations in India are not confined to the family planning program; indeed, the problems of federalism seem to influence almost all national programs that require high-level cooperation between the center and the states. Some observers have noted a deterioration in the ability of the center to carry through its objectives at the level of the states. Paul Brass, a respected student of Indian politics, has recently written, "Despite strong centralizing drives by [Indian National] Congress governments in Delhi . . . there have been recurring problems in center–state relations and long term trends that favor regionalism, pluralism and decentralization." Brass points out that in several important policy areas in which the states hold sole or primary constitutional authority, they are able, by their actions and nonactions, to "prevent the adoption of uniform policies for the country which the national leadership considers essential for the general processes of economic growth, development, and social justice" (Brass, 1990: 60).

The changes that have been taking place in center–state relations in India not only represent a shift in power from the center to the states, but also reflect a change in the character of political leadership, especially in the states. In the early years of independence, political and administrative leaders were drawn, in the main, from among the Westernized elite: highly educated, cosmopolitan in outlook, and socialized to western political and bureaucratic norms; today's leaders, by contrast, frequently received their education in a local language, have closer ties to local communities, and seem more responsive to local demands. While political and administrative sensitivity to local needs and demands is desirable, the absence of demand for family planning may deprive the program of state and local funds and commitment. Local political control may also result, as has frequently occurred at the district level in India, in unwarranted political interference in the delivery of family planning services. For example, several authors have commented that political patronage is often employed to influence such decisions as the use of vehicles, the location of clinics, appointments, promotions, and disciplinary actions (Maru, 1990; Bhatt, 1987). The extension of political responsiveness at the state and local levels may therefore come at the cost of the national capacity to implement important development programs—a problem that is by no means unique to India.

## Transnational actors and family planning programs

Previous sections of this introduction have emphasized the contributions of governmental foreign assistance programs and official multilateral agencies to the development—as well as the politics—of organized family planning efforts in the third world. While the actions of these agencies are by definition transnational in character,[18] it should not be forgotten that transnational actors in the private sector have also been exceptionally influential in the family planning arena. Such American and European organizations as the International Planned Parenthood Federation, the Population Council, the Ford and Rockefeller Foundations, and the Pathfinder Fund were among the first and most significant promoters of international population assistance, especially in the period prior to the entry of governments and the United Nations system.

In recent years, the politics of family planning has been enlivened by the entry of new transnational actors into the arena.[19] While some such actors, for example the diffuse collectivity of biologists, ecologists, and others that constitutes the environmental movement, are voicing their support for the expansion of family planning programs, others are more closely involved in detailed attempts to influence the design and implementation of the programs. By far the most significant organization in the latter category is the Catholic Church, which, while generally tolerating the existence of family planning programs, has gathered its forces in an effort to roll back the spread of legalized abortion and sterilization (Paige, 1983; Crane, this volume). Another critical voice that has emerged more recently is that of the international feminist movement. Although internally divided on many issues, feminists have subjected family planning programs as currently constituted to a thoroughgoing and at times severe critique. In this last section, we will discuss these two transnational political actors.

Much of the fervor and controversy involving the politics of population in recent years stems from the intertwining of the politics of family planning with the politics of abortion. In large part this reflects the exporting of a political debate from the industrial countries, chiefly the United States, to the countries of the third world. The growing conservatism of

Washington and Rome in the 1980s regarding population issues set the tone for a heightened debate in many countries between groups in favor of, and those opposed to, abortion rights. Anti-abortion movements in the United States and Europe have assisted in the development of comparable groups in many developing countries; and similarly, prochoice, family planning, and feminist groups have encouraged and supported third world women who are working for abortion rights in their own countries. At present, efforts to liberalize abortion in developing countries where it is tightly restricted, or at least nominally outlawed, have generally yielded little success through open political debate,[20] but this is a potentially explosive battleground for the future.

## The Catholic Church

It is not easy to unravel the network of channels, formal and informal, direct and indirect, through which the Church may bring its influence to bear on governmental policies around the world. While the Church is the supreme moral and spiritual guide for millions of Catholics, it may also be regarded as a political organization with many ways of influencing the political decisionmaking of governments and international organizations. The Church's diplomatic missions, its national episcopal conferences, and other formal organs in Rome and in individual countries, provide the Vatican with direct links to governmental and national leaders at the highest levels.[21] Official Church documents, the scholarly and popular Catholic press, the pastoral letters of bishops, and the many views expressed by Church leaders and clergy on committees and commissions are but a few of the vehicles used to disseminate the official views of the Church indirectly to policymakers as well as to individual Catholics.[22]

In its two thousand years of existence, the Church has evolved into a complex, decentralized bureaucracy that speaks with many voices carrying different degrees of authority (Vallier, 1973; Keely, this volume). The diversity of opinion within the Church was accentuated by the Second Vatican Council, held in the early 1960s, which ushered in an era of greater decentralization and encouraged national hierarchies to take more initiatives. At the same time, priests and Church officers were urged to have "continuous dialogue with the laity" (Maguire, 1983: 805). Vatican II also asserted the right of the Church "to pass moral judgments, even on matters touching the political order, whenever basic personal rights or the salvation of souls make such judgment necessary" (US Catholic Conference, 1976; cited in Paige, 1983: 53–54). In effect, the Vatican Council loosened its hold over both the national churches and the laity and implicitly invited Catholics to engage in political actions on behalf of the poor or disadvantaged, regardless of whether the latter were Catholics.

The Church's official position on contraception has been articulated during this century in a series of documents emanating from the Vatican. The most authoritative of these are the two papal encyclicals, *Casti connubi*, which in 1930 contained the first official condemnation of contraception (Donaldson, 1988), and *Humanae Vitae*, the encyclical of Pope Paul VI, which in essence reaffirmed the conservative position taken by *Casti connubi*. The publication of *Humanae Vitae*, in 1968, was a disappointment to many Catholics, clergy and laity alike, who had hoped that the social liberalism espoused by the Second Vatican Council might be extended to the domain of sexuality and reproduction and lead to a softening of the official position on contraception, if not abortion (Paige, 1983, including citations and footnotes therein). Some Church authorities have in fact tended to soft-pedal their positions on contraception (Keely, this volume), but at its core the teachings of *Humanae Vitae* remain in place. Nevertheless, the Church has exercised caution, flexibility, and diplomatic skill in drafting its official pronouncements on family planning. At times, as in the Holy See's official statement to the United Nations Conference on Environment and Development at Rio de Janeiro (Holy See, 1992), the Church has seemed to support family planning with only minor caveats; at other times, the Church has taken advantage of more favorable circumstances—in countries with weak governments, prominent Catholic politicians and bureaucrats, and strong Church leaders, for example—to state less ambiguously its opposition to artificial methods of contraception.

In contrast to its implicit position on contraception, the Church's opposition to abortion has not wavered since the late nineteenth century. Although a number of leading Jesuit thinkers and other theologians have questioned whether abortion should always be thought of as an act of homicide (Maguire, 1983), the political influence of the Church and of many individual Catholics on abortion and sterilization in recent years has become more conservative. In both developed and less developed countries, public debate over the possible legalization of abortion, or the inclusion of sterilization as a method of family planning, has often been the occasion for the reaffirmation of Church orthodoxy. For example, the design of the new family planning program in Peru was changed to exclude sterilization and abortion as a direct result of pressures exerted by the Catholic Church (Aramburú, this volume).

The increasingly frequent articulation of official Catholic doctrine on abortion, sterilization, and divorce during the past decade or more has been only part of a broader campaign intended to bring peripheral units of the Church back under central control and to restore doctrinal orthodoxy and discipline (Keely, this volume). The appointment of the ultraorthodox Cardinal Joseph Ratzinger as head of the Sacred Congregation for the Doctrine of the Faith, the committee charged with assuring conformity on doctrinal matters; the ban on liberation theology, which combines Christian beliefs with Marxist analysis; and the silencing of other dissenting theologians are only some of the more significant actions taken by Pope John Paul II to effect this agenda (see, for example, *The Washington Post,* 1985a and 1985b; *The New York Times,* 1986 and 1990). Within this domain, moreover, John Paul II's personal espousal of traditional doctrine on matters of sexuality, reproduction, and the family has done much to spread conservative orthodoxy, especially in the countries of Africa and Latin America where the majority of Catholics reside.

It is important not to confuse the official views and actions of the Church with those of Catholics acting on their own or in association with others who share their convictions. These lines of distinction are not always clearly defined, however. For example, because the National Right-to-Life Committee (NRLC) movement in the United States counts many Catholics among its members, it is widely perceived to be an official Church organization. It is not always realized that, while this was true in the 1960s and early 1970s, the Church severed its formal connection with the NRLC after a number of fundamentalist Protestant groups joined and radicalized it (Paige, 1983). There may also be a blurring of the lines between the Church and the layman's group, Opus Dei. This highly conservative organization, which works internationally to promote official doctrine—on reproductive as well as other issues—has the strong personal support of Pope John Paul II (*The New York Times,* 1992), but is not an official organ of the Church. Indeed, Catholic clergy are involved in many social and political organizations, but these activities should not be confused with official Church doctrines and policies. From the perspective of politicians and administrators formulating population policies, however, there may be little difference between official and unofficial Catholic activities, especially in countries where the political system is not highly developed and the Church is one of the few competing centers of power.

The pronouncements and actions of the Church on reproductive and family issues may easily be interpreted as signifying that the Church is engaged in a struggle against forces—feminists and prochoice groups, for example—that are external to the Church itself. While not incorrect, this interpretation neglects an important aspect of the controversy. The Church as a complex bureaucratic organization encompasses numerous internal factions with differing points of view on these and other issues; however, internal differences are much less likely to be made public than are external debates. The conservative position of the Church on family and sexual issues today is closely identified with Pope John Paul II and is related to his broader objective to still dissenting voices and reestablish central control within the Church. Many knowledgeable individuals, however, feel that the present conservatism on sexual matters may not portend the long-term future. They believe that in days to come, as in the past, the Church will accommodate itself to the changes in its social environment and will become more sympathetic to the problems of global population growth and even to the realities of women's lives.

### The feminist movement

In recent years, a new voice of growing strength and influence—the voice of women organized to defend and advance the interests of women—has started to be heard in family planning circles. Encouraged in part by the activities of the United Nations Decade of Women, 1975–85, as well as by activities sponsored by UNFPA, the Population Council, Pathfinder International, and the development agencies of the Nordic countries, among others, third world women's groups have begun to exercise increasing, although still limited, influence over the implementation of family planning programs. Significantly, many third world women's groups have forged links with international coalitions of women—often initiated by activist women's groups in the West—through which third world women's voices are magnified and from whom they can receive support and assistance. Not all women's groups have addressed themselves to issues of reproductive health, but those that have are providing a new and searching critique of orthodox family planning programs (Dixon-Mueller and Germain; Crane; both in this volume).

In the main, there is a high level of agreement among women's groups that women should have the right to make informed, unconstrained choices on reproduction and to have free access to high-quality family planning services. As Rosalind Petchesky reports, by the time of the United Nations Conference on the Decade of Women, held in Nairobi in 1985, "[T]he promotion of reproductive rights as fundamental to women's achievement of a just status in society had become a worldwide goal of women's rights activists" (Petchesky, 1990: 1). Despite the underlying agreement that the availability of family planning is central to women's status and welfare, feminist groups differ markedly among themselves and with orthodox family planning programs, both on the definition of reproductive rights and the means by which they should be attained. Many feminists are concerned by what they see as a growing emphasis in family planning programs on the promotion of what are commonly considered, in demographic terms, more effective methods—IUDs, sterilization, and such long-acting hormonal methods as injectables and implants. Their critique is twofold: that such methods are less easily reversed than are simpler barrier methods and thus reduce women's control, and that their invasiveness poses greater risks to women's health.

Some third world women's groups have already demonstrated their ability to influence policy and programmatic decisions as they relate to women's health and, in particular, to the range and types of contraceptives that are offered by family planning programs. In 1989, a Peruvian feminist organization, Movimiento Manuela Ramos, organized a public campaign of opposition to a new policy strategy proposed by USAID, and succeeded in getting the language of the proposal changed (Petchesky and Weiner, 1990). The issues addressed were USAID's proposal to subsidize only long-term methods—IUDs, sterilization, and implants—and to remove the subsidies from all other methods on the grounds that these were readily available in the private sector. Opposition was also directed to what was seen as a de-emphasis by USAID on family planning information and counseling. In Brazil, feminist groups succeeded in ending the clinical trials of NORPLANT® by challenging the safety and convenience of the method itself, as well as what they saw as inadequacies in the research protocols (Barroso and Correa, 1991). In the Phillipines feminists contrived to get limited family planning services reintroduced after the program had been dismantled by Corazon Aquino's government (Dixon-Mueller and Germain, this volume).

While informed choice, safe contraceptive technology, and high quality of care are issues on which women's groups can generally agree, there is considerable disagreement on other aspects of population and family planning policy. A number of feminist groups have rejected the demographic rationale as an unacceptable foundation for family planning programs, arguing that it subordinates the interests of women to an abstract societal good. They also contend that exploitation of the poor by the rich, rather than population growth, is the true cause of social ills. However, some individual feminists seem increasingly inclined to take a less intransigent position. Carmen Barroso (1990), in a paper presented to a conference of the Women's Global Network for Reproductive Rights, argues that the existence of a demographic policy is immaterial provided that the content of the policy is consonant with such feminist objectives as freedom of choice and women's reproductive health. Others go further, arguing that feminists must confront the joint realities that the world cannot sustain an unlimited population, and that high rates of illegal abortion, sterilization, and acceptance of any available method of contraception by third world women indicate that there are high levels of unwanted pregnancies. Berer (1991) argues that feminists must develop the concept of a feminist population policy or risk being isolated and ignored in the ongoing international debate over population policy.

Although the intervention of feminist groups in family planning policy decisions has taken place in only a few countries so far, these cases should be seen as intimations of what is likely to become a much greater feminist presence in the years ahead. The feminist agenda is broad, encompassing questions concerning the legitimacy of population policies and the legalization of abortion (Crane, this volume), as well as more programmatic issues such as quality of care, access, informed consent, and control by women, among others. The growing number of feminist groups in third world countries, and their collegial as well as tutelary links with feminist organizations in developed countries, provide women's organizations with numerous ways to influence family planning programs. Pressures can be brought to bear indirectly through foreign donors as well as directly on policymakers and program officials at national, regional, and local levels in developing countries. Hitherto, the impact of the feminist movement in family planning has been concentrated in the countries of the Western Hemisphere, and may be starting in some East and Southeast Asian countries. Women's groups in Africa, the Middle East, and South Asia generally have further to go in overcoming social, cultural, and political barriers to achieving political influence, although there are indications that the salience of fertility questions to women in these regions may enable them to exercise influence on the design and implementation of population programs in the future.

• • •

Thirty years ago, the population debate encompassed three main positions, each of which grew out of a particular political orientation or philosophy. Subscribers to these three positions may be loosely defined as: (1) those economists and economic demographers who argued the need for a reduction in the rate of population growth in order to remove a major impediment to development; (2) the Soviet Union, its allies, and Marxist ideologues in general, who held that population trends are a product of economic and social relations; and (3) the Catholic Church, which at that time was in a liberalizing phase stimulated by the Second Vatican Council. In the intervening period, the positions of all three sets of actors have evolved. While maintaining their conviction that it is necessary for the world to reduce its rate of population growth, the "population controllers," to use a convenient shorthand, have toned down their "crisis" approach and have adopted a more moderate stance. The collapse of the Soviet Union and the worldwide discrediting of Marxist ideology have reinforced Marxist demographers' gradual recognition that population change follows its own internal dynamics. Finally, the Catholic Church has turned the clock back and is attempting to restore Church discipline, recentralize decisionmaking, and, at the same time, reaffirm traditional doctrine on the use of artificial methods of contraception. This conservative trend has stimulated political controversy both within and outside the Church.

While the administration of policy in most fields tends to become somewhat routine after 30 years of continuous implementation, the same cannot be said of population policy. The emergence of new global concerns, new perspectives, and new actors with interests related to family planning programs and policies has served to keep the level of controversy at a high pitch, both globally and within individual states and localities. To some extent, the continuity of, and even the increase in, the amount and variability of political debate in this area is a function of growing experience and the proliferation of knowledge related to population change and family planning. In part, it arises from the conservative trend that is apparent in the Church and some donor countries. In many ways, however, the greater political activism we witness in developing countries is a concomitant of development itself, reflecting higher levels of education, more effective emancipation, and a growing sense of confidence among women that they can take control of their own lives. The final irony is that the same spirit of political activism that grew out of development should be used to attack family planning programs that have encouraged development itself to take place.

## Outline of the book

The articles in this volume are divided into four sections: overviews and frameworks for analysis; the political environment of policies and programs to influence family planning; case studies of the politics of policy formulation and implementation; and transnational actors and family planning policy.

In the first section Joseph Chamie draws on the United Nations Monitoring Reports to analyze trends in population policy among the countries of the world and to identify some of the difficulties inherent in studies based solely on official data. This is followed by John Thomas and Merilee Gindle's presentation of a framework for the analysis of policy change, focusing on policy elites, the social, political, economic, and administrative context within which they operate, the circumstances that surround specific policy issues, and the characteristics of the policy itself. The authors find not only that population policymaking is more "political," but also that it differs in other respects from policymaking in other development fields.

The second section contains four articles that analyze the political environment of population and family planning policy in specific regions and countries. Omari Kokoli, on sub-Saharan Africa, focuses on ethnic diversity and the legacy of the colonial experience. V. A. Pai Panandiker and P. K. Umashankar discuss the role of population numbers in electoral politics at the central and state levels in India. This is followed by Gustavo Cabrera's study of the evolution of population policy in a corporate society, Mexico. Finally, Ali Mazrui discusses the influence of Islam on fertility and family planning in Africa.

Following these are three case studies of the politics of population policy and family planning. Tyrene White analyzes the relationship between economic and population policy in the People's Republic of China. In his analysis of the relationship between population policy and fertility decline in Latin America, Carlos Aramburú asks whether population policy is a necessary prelude to fertility change. Donald Warwick focuses on the role of politics in shaping what is and is not researched in the population field.

The volume continues with three articles on the role of transnational actors in the formulation and implementation of family planning programs. First, Ruth Dixon-Mueller and Adrienne Germain discuss the extent and influence of women's political action on family planning in developing countries, with particular reference to Brazil, Nigeria, and the Philippines. Charles Keely analyzes the limits of Papal power since *Humanae Vitae*. Barbara Crane discusses the transnational politics of abortion.

The volume concludes with a brief assessment of the future issues in this field of study.

## Notes

The authors thank Scott Grosse, Barbara Crane, and Joseph Chamie for their insightful comments.

1 "Overpopulation" has received periodic attention, especially from British scholars. The most widely discussed episode was that associated with the start of the industrial revolution, the occasion in 1789 for Malthus's famous *Essay on the Principle of Population* (Malthus, 1976).

2 In this essay, the term "family planning" is most often used to denote family planning policy or program implementation. The term may also be used, of course, to denote the decision of individual couples to plan the number and timing of births.

3 These authors are not suggesting that there is no relationship between population and national power, but rather that technical and economic superiority on which national power is now based is mediated by the quality of the population, as well as by such factors as the structure of a nation's alliances. Some authors have argued that if the levels of technological progress and economic productivity were held constant, the country with the largest population would have the advantage (see Organski, Bueno des Mesquita, and Lamborn, 1972; Wright, 1958).

4 The low-level equilibrium trap refers to the situation in which countries with underdeveloped economies, a stable equilibrium level of income per capita, and growing populations would be unable to increase the level of investment in capital equipment per worker. Thus, the economy would be unable to grow. Harvey Leibenstein (1954) developed a related economic model at about the same time.

5 India substantially increased funding for family planning after the publication of its 1960 census. Likewise, after the advent of General Ayub Khan as president in 1958, Pakistan started to allocate significant funds to family planning, especially in the Second Five Year Plan, 1961–65. Ceylon adopted a policy in 1965, although it had earlier received official assistance from Sweden in introducing family planning into the government health service. These countries were followed by Tunisia (1961), Malaysia, Mauritius, China, Iran, Kenya, Singapore, Turkey, Barbados, and Nicaragua (1965–67), and Indonesia, Morocco, Ghana, Taiwan, Jamaica, and Trinidad and Tobago (1968–69). (See Myrdal, 1968: 1489–1494; Nortman, 1974, Table 8.)

6 Richard Gardner, the United States delegate to the 1962 General Assembly, argued that the United Nations already had the necessary authority to provide technical assistance for the formulation and execution of population policies—an opinion that was evidently found to be legally sound (see Symonds and Carder, 1973). However, the United States did not start to provide technical assistance until President Johnson assumed office.

7 The lack of central control was under discussion frequently in the United Nations during the 1960s, prompting the commissioning of an official report, *Study of the Capacity of the UN Development System* (the Jackson Report), in 1969. In this document, Sir Robert Jackson used a biological metaphor to de-

scribe the United Nations development system as "a system without a brain" (Symonds and Carder, 1973: xii and 192). The present Secretary-General, Boutros Boutros-Ghali, is aware of the cost of duplication and overlapping mandates and is said to be attempting to restructure the system to achieve better coordination.

8 The establishment by Franklin D. Roosevelt of new agencies to implement his New Deal policies has been interpreted as intended to ensure that his programs had an opportunity to become strong and autonomous before they were folded into existing sector agencies (Rourke, 1976; see also Schlesinger, 1959).

9 John Waterbury (1972) argued that Egypt, as well as other Arab countries in the 1950s and 1960s, saw their population problem as essentially one of maldistribution. The solution was often thought to lie in irrigating the deserts and resettling the population on the newly fertile land. Reflecting later, Waterbury characterized Revolutionary rule in Egypt as having often "involved a search for solutions to real problems by stepping outside their parameters." An example of this type of response is:

> . . . to treat knotty problems as insoluble and end-run them. Thus if the challenge of changing the behavior of 5 million peasants is too awesome, one may still modernize agricultural production by farming the peasantless desert and by transplanting people to a new way of life. Or if modifying fertility behavior contains unacceptable costs among a conservative Muslim population, one can ponder the possibility of large-scale population transfers to the rich agricultural voids of Syria, Iraq, or the Sudan. (Waterbury, 1983: 49)

10 Warwick's analysis is consonant with an earlier assessment by Gayl Ness (1971) that ends soon after the government policy was adopted.

11 Organization theorists have long stressed the necessity for organizations to seek additional funds, new programs, and strong alliances in order to grow, adapt to changing environments, and, ultimately, to survive (see Barnard, 1946; Drucker, 1958; Simon, Smithburg, and Thompson, 1950).

12 In the early 1970s WHO objected to a UNFPA/World Bank project in Indonesia on the grounds that it would drain physicians and other scarce resources that would otherwise belong to the health system (see Finkle and Crane, 1976).

13 Although family planning has gained more legitimacy among ministries of health since it has been accepted as one of the basic components of primary health care, observation in the field suggests that it may be less vigorously promoted than are other components, perhaps because doing so is more difficult and requires more time and effort to motivate clients.

14 An article in the *Jerusalem Post* (4 April 1990) cites a "well-informed foreign observer who had been told of an Egyptian imam who told his flock, ' I am required by the government to tell you that for the good of the country and for your own good you should limit the size of your families. So I have told you. But both you and I know the truth is the exact opposite.' "

15 For an excellent discussion of the political and organizational difficulties of implementing the family planning program in Uttar Pradesh in the early 1970s, see Simmons and Ashraf, 1978: 22–34.

16 To take one example, it has been extremely difficult to move the program from its reliance on sterilization to the use of a broader range of methods (Basu, 1984). Yet it has been shown elsewhere that acceptance and continuation rates rise significantly with each additional method provided (Jain, 1989; Phillips et al., 1982).

17 A recent analysis shows that India's total fertility rate declined by only 1.06 births per woman between 1960–64 and 1980–84. However, approximately 75 percent of this decline is attributable to declining marital fertility (Retherford and Rele, 1989).

18 That is, while based in one country, these agencies have policies, programs, and other activities that relate to or take place in other countries. Examples of transnational actors include multinational organizations, some foundations, religious organizations that have an international following, and some development and/or relief agencies.

19 For further discussion of the impact of new transnational actors, including the environmental movement, on national and international family planning policies and programs, see Crane (1993).

20 An exception is Botswana where, after an intense open debate, abortion was legalized in 1991 (see Botswana, National Assembly, 1991, Section 160). The intensity and openness of the debate are reflected in the many full- and half-page reports published in *The Botswana Guardian* and *Mmegi Reporter* (Gabarone) between May and September 1991.

21 Examples abound of such influence being brought to bear in the area of reproductive health. The influence of Cardinal Sin and the Church in the Philippines on the dismantling of the family planning program by President Aquino may be the best-known incident in recent years (Clad, 1988). William Wilson, the first United States ambassador to the Vatican, has recently claimed that diplomatic activity between the Vatican and the White House was a strong influence on the United States' "Mexico City policy" (see Bernstein, 1992). It has been reported that the Vatican has engaged in similar diplomatic activity with national delegations to the United Nations Conference on Environment and Development held in Rio de Janiero (*The New York Times*, 28 May 1992).

22 For example, the Permanent Council of Catholic Bishops in 1979, the year in which the French abortion law came up for review and permanent enactment, issued a White Book setting out the official position of the Church. The book was published and distributed in bookshops. It was also widely discussed in the press. Similarly, the issuance of a Pastoral Letter by the Conference of [West] German Bishops, attacking the Social Democrat/Liberal coalition government for its position on abortion and divorce on the eve of the 1980 parliamentary elections, was widely interpreted as an instruction to Catholics about how to vote (McIntosh, 1983).

## References

Adepoju, Aderanti. 1981. "Military rule and population issues in Nigeria," *African Affairs* 80: 318.

Balfour, Marshall. 1962. "Chairman's report of a panel discussion on the comparative acceptability of different methods of contraception," in *Research in Family Planning*, ed. Clyde V. Kiser. Princeton: Princeton University Press, pp. 373–386.

Barnard, Chester. 1946. *The Functions of the Executive*. Cambridge, MA: Harvard University Press.

Barroso, Carmen. 1990. "Maternal mortality: A political question," in *Maternal Mortality and Morbidity: A Call to Women for Action*. Amsterdam: Women's Global Network for Reproductive Rights and Latin and Caribbean Women's Health Network, pp. 4–5.

———, and Sonia Correa. 1991. "Servidores públicos versus profesionales liberales: la política de la investigación sobre anticoncepción" [Public servants versus professional liberals: The policy for contraceptive investigation], *Estudios Sociologicos* 9, 25: 75–104.

Basu, Alaka Malwade. 1984. "Ignorance of family planning methods in India: An important constraint on use," *Studies in Family Planning* 15, 3: 136–142.

Berer, Marge. 1991. "What would a feminist population policy be like?" *Conscience* 12, 5: 1–5.

Bernstein, Carl. 1992. "The holy alliance," *Time*, February 24: 28–35.

Besemeres, John F. 1980. *Socialist Population Politics: The Political Implications of Demographic Trends in the USSR and Eastern Europe*. White Plains, NY: M.E. Sharpe.

Bhatt, Anil. 1987. "The social and political dimensions of administering development in India," in *Beyond Bureaucracy: Strategic Management of Social Development*, eds. John C. Ickis, Edilberto de Jesus, and Rushikesh Maru. West Hartford, CT: Kumarian Press, pp. 102–115.

Botswana, National Assembly. 1991. Penal Code (Amendment) Act. 20 September, 1991.

Brass, Paul. 1990. *The Politics of India Since Independence*. New Cambridge History of India. Cambridge/New York: Cambridge University Press.

Caldwell, John C., and Pat Caldwell. 1987. "The cultural context of high fertility in sub-Saharan Africa," *Population and Development Review* 13, 3: 409–437.

Chamie, Joseph. 1981. *Religion and Fertility: Arab Christian–Muslim Differentials*. New York: Cambridge University Press.

Clad, James. 1988. "Genesis of despair," *Far East Economic Review* 20 (October): 24–28.

Cline, Ray S. 1975. *World Power Assessment: A Calculus of Strategic Drift*. Boulder, CO: Westview Press.

Coale, Ansley J., and Edgar M. Hoover. 1958. *Population Growth and Economic Development in Low-income Countries: A Case Study of India's Prospects*. Princeton: Princeton University Press.

Cox, Robert W., and Harold K. Jacobson. 1973. *The Anatomy of Influence: Decision Making in International Organizations*. New Haven: Yale University Press.

Crane, Barbara B. 1993. "International population institutions: Adaptation to a changing world order," in *Institutions for the Earth: Sources of Effective International Environmental Protection*, eds. Peter M. Haas, Robert O. Keohane, and Marc A. Levy. Cambridge, MA.: M.I.T. Press, pp. 351–393.

Dixon-Mueller, Ruth. 1987. "United States international population policy and 'the woman question,'" *Journal of International Law and Politics* 20, 1: 143–168.

Donaldson, Peter J. 1988. "American Catholicism and the international family planning movement," *Population Studies* 42, 3: 367–373.

Drucker, Peter F. 1958. Business objectives and survival needs: Notes on a discipline of business enterprise," *Journal of Business* 31: 81–99.

Dyson, Tim, and Mick Moore. 1983. "On kinship structure, female autonomy, and demographic behavior in India," *Population and Development Review* 9, 1: 35–60.

Ekanem, I. I. 1972. *The 1963 Nigerian Census: A Critical Appraisal*. Benin City, Nigeria: Ethiopia Publishing.

Eversley, David E. C. 1959. *Social Theories of Fertility and the Malthusian Debate*. Oxford: Clarendon Press.

Faour, Muhammad A. 1989. "Institutional constraints to family planning in the Arab East." Paper prepared for the seminar on The Role of Family Planning Programs as a Fertility Determinant, sponsored by the International Union for the Scientific Study of Population, the Population Council, the Rockefeller Foundation, and the Population Research Unit of the League of Arab States, Tunis, 26–30 June.

Finkle, Jason L. 1972. "The political environment of population control in India and Pakistan," in *Political Science in Population Studies*, eds. Richard L. Clinton, William S. Flash, and R. Kenneth Godwin. Lexington, MA.: D.C. Heath, pp. 101–128.

———. 1982. "Budgets, bureaucracy and family planning: Some programmatic implications of non-budgeting," in *Public and Private Expenditures in the Health Care Sector of Egypt*, eds. Robert Grosse, Demetrius Plessas, and Helmi El Bermawy. Ann Arbor: University of Michigan, Department of Population Planning. Mimeo.

———, and Barbara B. Crane. 1976. "The World Health Organization and the population issue: Organizational values in the United Nations," *Population and Development Review* 2, 3/4: 367–393.

———, and C. Alison McIntosh. 1980. "Policy responses to population stagnation in developed societies," in *Social, Economic and Health Aspects of Low Fertility*, ed. Arthur A. Campbell. Washington, DC: United States Department of Health, Education and Welfare, National Institutes of Health. United States Government Printing Office.

Gadalla, Saad. 1979. "Egypt: Cultural values and population policies." Social Research Center, American University of Cairo. Unpublished manuscript.

Germain, Adrienne. 1987. "Reproductive health and dignity: Choices by Third World women." Paper presented at the International Conference on Better Health for Women and Children through Family Planning, Nairobi, Kenya. New York: International Women's Health Coalition.

Gille, Halvor. 1961. "What the Asian censuses reveal," *Far Eastern Economic Review* 29 (June): 635–641.

Gilmour, David. 1983. *Lebanon: The Fractured Country*. Oxford: Martin Robertson.

Glass, David V. 1940. *Population Policies and Movements in Europe*. Oxford: Clarendon Press.

Gold, Rachel Benson, and Sandra Guardado. 1988. "Public funding of family planning, sterilization, and abortion services," *Family Planning Perspectives* 20, 5: 228–233.

Greenhalgh, Susan. 1993. "The peasantization of the one-child policy in Shaanxi," in *Chinese Families in the Post-Mao Era*, eds. Deborah Davis and Stevan Harrell. Berkeley: University of California Press, pp. 219–250.

Hartmann, Betsy. 1987. *Reproductive Rights and Wrongs*. New York: Harper and Row.

Holy See. 1992. Statement of Archbishop Renato R. Martino. Apostolic Nuncio, Head of the Holy See Delegation to the United Nations Conference on Environment and Development. Rio de Janeiro, 4 June, 1992.

Jain, Anrudh K. 1989. "Fertility reduction and the quality of family planning services," *Studies in Family Planning* 20, 1: 1–16.

King, Preston. I. 1982. *Federalism and Federation*. London: Croom Helm.

Kirk-Greene, A. H. M. 1971. *Crisis and Conflict in Nigeria; A Documentary Sourcebook, 1966–1970*. London: Oxford University Press.

Knarr, John R. 1976. "Population politics and the Soviet polity." Ph.D. dissertation, University of Los Angeles.

Leibenstein, Harvey. 1954. *A Theory of Economic-Demographic Development*. Princeton: Princeton University Press.

Maguire, Daniel C. 1983. "Abortion: A question of Catholic honesty," *The Christian Century* (September): 14–21.

Malthus, Thomas Robert. 1976. *An Essay on the Principle of Population*, ed. Philip Appleman. New York: W. W. Norton.

Maru, Rushikesh. 1990. "Politics and family planning implementation at the field level in India." Paper presented at the seminar on the Politics of Induced Fertility Change. Bellagio, Italy, 19–23 February.

Mazrui, Ali A. 1971. "Public opinion and the politics of family planning," *Rural Africana* 14 (Spring): 38–52.

McIntosh, C. Alison. 1983. *Population Policy in Western Europe: Responses to Low Fertility in France, Sweden and West Germany*. Armonk, NY: M.E. Sharpe.

Miller, Norman N. 1971. "The politics of population." American Universities Field Staff Reports, East Africa Series 10, 2 (Kenya).

Mosher, William D., and Calvin Goldscheider. 1984. "Contraceptive patterns of religious and racial groups in the United States, 1955–76: Convergence and distinctiveness," *Studies in Family Planning* 15, 3: 101–111.

Myrdal, Gunnar. 1968. *Asian Drama*. Vol. 2. New York: Random House.

Nelson, R. R. 1956. "A theory of the low-level equilibrium trap," *American Economic Review* (December): 894–908.

Ness, Gayl D. 1979. "Organizational issues in international population assistance," in *World Population and Development: Challenges and Prospects*, ed. Philip M. Hauser. Syracuse: Syracuse University Press, pp. 615–649.

———. 1993. "The powers and limits of state and technology: Rice and population in South-East Asia," in *Population-Environment Dynamics: Ideas and Observations*, eds. Gayl D. Ness, Steven R. Brechin, and William D. Drake. Ann Arbor: University of Michigan Press, pp. 109–132.

———, and Hirofumi Ando, 1971. "The politics of population planning in Malaysia and the Philippines," *Journal of Comparative Administration* 3: 296–329.

*The New York Times*, 1986. "Vatican orders a theologian to retract teachings on sex." 11 March.

———. 1990. "The Vatican warns Catholic theologians over public dissent." 27 June.

———. 1992. "Priest challenges sainthood moves." 13 January.

Nortman, Dorothy. 1974. *Population and Family Planning Programs: A Factbook*. Reports on Population/Family Planning, no. 2. New York: The Population Council.

Organski, A. F. K., Bruce Bueno des Mesquita, and Allen Lamborn. 1972. "The effective population in international politics." United States Commission on Population Growth and the American Future. Research Reports, vol. 4, *Governance and Population*. Washington, DC: United States Government Printing Office, pp. 235–250.

Overbeek, Johannes. 1974. *History of Population Theories*. Rotterdam: Rotterdam University Press.

Paige, Connie. 1983. *The Right-to-Lifers: Who They Are, How They Operate, Where They Get Their Money*. New York: Summit Books.

Petchesky, Rosalind P. 1990. "Feminist perspectives on reproductive rights since 1985," in Petchesky and Weiner, 1990, pp. 1–9.

———, and Jennifer A. Weiner. 1990. *Global Feminist Perspectives on Reproductive Rights and Reproductive Health: A Report on the Special Sessions held at the Fourth International Interdisciplinary Congress on Women, Hunter College, New York City*. 3–7 June, mimeo.

Phillips, James F., et al. 1982. "The demographic impact of the Family Planning–Health Services Project in Matlab, Bangladesh," *Studies in Family Planning* 13, 5: 131–140

———, Ruth Simmons, and Marjorie A. Koblinsky. 1985. "Bureaucratic transition: A paradigm for policy development in Bangladesh." Paper prepared for the seminar on Societal Influences on Family Planning Program Performance. International Union for the Scientific Study of Population. Jamaica, 10–13 April.

Piotrow, Phyllis T. 1973. *World Population Crisis: The United States Response*. New York: Praeger Publishers.

Retherford, Robert D., and J.R. Rele. 1989. "A decomposition of recent fertility changes in South Asia," *Population and Development Review* 15, 4: 739–747.

Rourke, Francis E. 1976. *Bureaucracy, Politics and Public Policy*. Boston: Little, Brown and Co.

Ryder, Norman B., and Charles F. Westoff. 1977. *The Contraceptive Revolution*. Princeton: Princeton University Press.

Salas, Rafael M. 1976. *People: An International Choice*. New York: Pergamon Press.

Sauvy, Alfred. 1963. *Malthus et les deux Marx: le problème de la faim et de la guerre dans le monde* (Malthus and the two Marxes: The problem of hunger and war in the world). Paris: Denoël.

Schlesinger, Arthur M., Jr. 1959. *The Age of Roosevelt*. Boston: Houghton Mifflin.

———. 1965. *A Thousand Days: John F. Kennedy in the White House*. Boston: Houghton Mifflin.

Schuman, Frederick L. 1948. *Soviet Politics at Home and Abroad*. New York: Alfred A. Knopf.

Selznik, Philip. 1949. *TVA and the Grassroots*. Berkeley: University of California Press.

Simon, Herbert A., D.W. Smithburg, and V. A. Thompson. 1950. *Public Administration*. New York: Alfred A. Knopf.

Simmons, Ruth S., and Ali Ashraf. 1978. "Implementing family planning in a ministry of health: Organizational barriers at the state and district levels," *Studies in Family Planning* 9, 2/3: 22–34.

Smith, Brian C. 1967. *Field Administration: An Aspect of Decentralization*. London: Routledge and Kegan Paul.

———. 1985. *Decentralization: The Territorial Dimension of the State*. London/Boston: George Allen and Unwin.

Strangeland, Charles Emil. 1904. "Pre-Malthusian doctrines of population: A study on the history of economic theory," in *Studies in History, Economics and Public Law* 2, no. 3. New York: Columbia University Press.

Symonds, Richard, and Michael Carder. 1973. *The United Nations and the Population Question, 1945–1970*. New York: McGraw-Hill.

United Nations Commission on International Development. 1969. *Partners in Development*. New York: McGraw-Hill.

United Nations Department of Economic and Social Affairs. 1973. *Determinants and Consequences of Population Trends*. New York: United Nations.

United Nations Development Programme. 1969. *Study of the Capacity of the United Nations Development System*. United Nations Document DP/5 1969. New York: United Nations.

United Nations Food and Agriculture Organization (FAO). 1959. Report of the 10th Session of the Conference. Resolution 13/59.

———. 1960. *Freedom from Hunger: Outline of a Campaign*. Rome: FAO.

Vallier, Ivan. 1973. "The Roman Catholic Church: A transnational actor," in *Transnational Relations and World Politics*, eds. Robert O. Keohane and Joseph S. Nye. Cambridge, MA: Harvard University Press, pp. 129–152.

Warwick, Donald P. 1982. *Bitter Pills: Population Policies and Their Implementation in Eight Developing Countries*. New York: Cambridge University Press.

———. 1986. "The Indonesian family planning program: Government influence and client choice," *Population and Development Review* 12, 3: 453–490.

*The Washington Post*. 1985a. "Pope in Venezuela condemns divorce." 28 January.

———. 1985b. "Pope focuses on liberation theology." 29 January.

Waterbury, John. 1972. "The Cairo workshop on land reclamation and settlement in the Arab world." American Universities Field Staff Reports. North Africa Series, 17, 1.

———. 1983. *The Egypt of Nasser and Sadat: The Political Economy of Two Regimes*. Princeton: Princeton University Press.

Westoff, Charles F., and Larry L. Bumpass. 1973. "The revolution in birth control practices of U.S. Roman Catholics," *Science* CLXXIX: 41–44.

Wildavsky, Aaron, and Jeffrey L. Pressman. 1979. *Implementation: How Great Expectations in Washington Are Dashed in Oakland*. Berkeley: University of California Press.

Wright, Quincy. 1955. *The Study of International Relations*. New York: Appleton-Century-Crofts.

———. 1958. "Population and United States foreign policy," in *Population and World Politics*, ed. Philip M. Hauser. Glencoe, IL: The Free Press, pp. 260–270.

# OVERVIEW AND FRAMEWORK FOR ANALYSIS

# Trends, Variations, and Contradictions in National Policies to Influence Fertility

JOSEPH CHAMIE

THE POPULATION POLICIES OF GOVERNMENTS are capable of changing rapidly and dramatically. Consider, for example, the following illustrative cases.

In November 1959, President Dwight D. Eisenhower of the United States said: "So long as I am President, this government will have nothing to do with birth control; this is something for private organizations to deal with." Fewer than ten years later, in August 1965, President Lyndon B. Johnson announced: "Let us in all our lands—including this land—face forthrightly the multiplying problems of our multiplying populations and seek the answers to this most profound challenge to the future of all the world."[1]

In 1962, during the first full-scale debate at the United Nations General Assembly on the population issue, the Netherlands maintained its traditional view that the State had no authority to interfere with the individual rights of its subjects. Throughout the discussions, the Dutch delegation sided with governments opposing references to "population policies" and "technical assistance" for national projects and programs. Three years later, the Dutch delegation joined with governments that recommended: (a) endorsing the expanded population program as proposed by the Population Commission of the United Nations; (b) inviting specialized agencies to consider the extension of their population activities; (c) making available additional funds for this purpose; and (d) directing the secretary-general to formulate a priority program.

In 1972, the President of Algeria stated in his Independence Day speech that "the increase of population is a long-term investment." The prevailing view at that time was that the development of Algeria's large land areas and its mineral resources required proportionately large human resources. Less than a decade later, the Algerian government modified its position, noting that rapid population increase had impeded both social and economic development, and it subsequently formulated a long-term policy to reduce fertility and control population growth.

The policy of the People's Republic of China in 1978 was that couples may have no more than two children. In 1979, that policy was changed: Couples were now obliged to have only a single child, except when given explicit permission to have another. No one would be allowed to have a third child (see Hardee-Cleaveland and Banister, 1988; Zeng, 1989).

An underlying hypothesis of this article is that the population policies of a country are not only a function of demographic phenomena, but are also the consequence of social, economic, and political conditions specific to the country. Population policies—and such shifts in policy as those noted above—are likely to be of paramount importance in determining governmental expenditures, programs, practices, and assistance, which in turn affect the well-being of individuals and influence demographic trends. Accordingly, it is essential for policymakers, planners, and researchers to have a thorough understanding of the population policies of governments. This chapter examines trends, variations, and contradictions in government views and policies that are directed at influencing fertility, which since World War II has been a central concern and focus of most governments' interventions in the area of population.

## Assessing fertility perceptions and policies

The perceptions and policies of governments toward fertility reported here are based on information from the Population Policy Data Bank, which is maintained by the Population Division of the Department of International Economic and Social Affairs (DIESA) of the United Nations. This source of information is particularly suitable for this subject for several reasons. First, because these policy data are truly global in nature. All members of the United Nations, which currently number 170 governments, are covered in the Policy Data Bank. (For a more detailed description of the data-collection process and methodology used in the United Nations monitoring of population policy, see United Nations, 1990b).

Second, the population policies of countries have been monitored on an ongoing and reasonably standardized basis since the mid-1970s. The results of this monitoring have been published in reports that have appeared on a biennial basis (United Nations, 1979, 1980, 1982, 1985, 1988a, and 1990a). From these reports, fertility perceptions and policies are available for six points during the period from 1976 to 1989.[2]

Third, and most valuable to policy analysts, a high degree of consistency has been maintained within the data over the 14-year period. The definitions, classification, and categories of the major variables relating to perceptions and policies toward fertility have remained essentially unchanged; thus, reasonably meaningful comparisons and contrasts among countries and regions both on a cross-sectional and longitudinal basis are made possible. For example, the proportion of countries that view their fertility levels as too high and that have policies aimed at reducing those levels can be determined. It is possible to specify the policy trends these countries have followed during the 1970s and 1980s and to consider the evolution of individual countries' policies.

These monitoring reports have also been distributed widely over this period, allowing the data to undergo extensive scrutiny. In addition to professional examination from within the population community, many policymakers and other government officials have had frequent opportunity to review the information in draft and published form, and to communicate omissions and errors to the United Nations Population Division. Such an evaluative feedback mechanism enhances the credibility of the reported data.

Nevertheless, the data under investigation also have shortcomings. One difficulty concerns the absence or ambiguity of national population policies. Determining the views of governments without clearly stated fertility policies within a comparative international framework is a risky undertaking that may result in biases and distortions. Fortunately, the number of such countries, particularly among the less developed regions, has declined since the early 1970s. In Africa, for example, many countries have formulated population policies fairly recently.

In a few cases, government ambiguity remains and fertility policies are open to interpretation. For example, the distribution and sale of contraceptives may be officially prohibited while they are, in fact, widely available. Elsewhere, laws may permit the sale and distribution of contraceptives, but supplies are limited or nonexistent. Some countries are undergoing significant changes in their policies, while others are involved in great internal conflict. For example, immediately after the overthrow of the Ceauşescu government in December 1989, Romania liberalized its policies on abortion, access to modern methods of birth control, and the reproductive rights of women. Interpretation of policies in ambiguous cases or in cases of turmoil may lead to an incomplete assessment or misrepresentation of fertility perceptions and policies.

Another issue, which has been raised at forums within and outside the United Nations, is the degree to which countries' policies, as reported by the Secretariat, are consistent with their efforts aimed at socioeconomic development. Are reported national fertility policies genuine reflections of governmental commitment, or are they largely, as some critics maintain, enunciated for public relations purposes or the securing of external funds? Furthermore, how are countries to be understood where government policy pronouncements differ greatly from government actions?

## Trends in fertility perceptions and policies

At the end of the last decade, the majority of countries in the world viewed their fertility levels as unsatisfactory; 44 percent indicated that their national levels were too high and 12 percent that they were too low. These countries generally adopted fertility policies that were consistent with their concerns, either to lower fertility levels that were believed to hamper development efforts, as in the cases of China, Egypt, India, and Mexico, or to raise them for the sake of development or political aims, as in the cases of France, Iraq, Israel, and Uruguay. The remaining 75 countries either report policies aimed at maintaining fertility at current levels, as in Czechoslovakia, Mali, and the former Soviet Union, or they do not intervene to influence fertility at all, as in Brazil, Canada, and Japan.

Significant shifts have taken place in the fertility perceptions and policies of governments since the mid-1970s. The proportion of countries viewing their fertility levels as too high, for example, increased from 35 to 44 percent between 1976 and 1989. Among the countries that shifted their positions are Algeria, Ethiopia, Jordan, Nigeria, Peru, Sierra Leone, Tanzania, and Zambia. During this period the proportion of governments that adopted policies to lower their fertility levels also jumped significantly, increasing from 25 percent to nearly 40 percent. The proportion of countries that regard their fertility levels as too low rose only slightly, from 12 to 14 percent in the same period.

To accompany these shifts, a growing number of countries reported direct governmental support for the provision of contraceptive services. By 1989 nearly three-fourths of the 170 countries provided such support, whereas in 1976 slightly less than two-thirds of the countries did so. Moreover, the number of governments limiting or restricting their citizens' access to contraceptive methods decreased from 15 to six over this period; the six governments are Cambodia, Iraq, Laos, Mongolia, Romania, and Saudi Arabia.[3]

Fertility levels in the developed world are generally at or below replacement level. The majority of countries in this category, including Australia, Canada, Japan, Netherlands, the former Soviet Union, Great Britain, and the United States, view their fertility levels as satisfactory and most follow policies of nonintervention. However, a significant minority of these countries—about one-fourth—view their fertility levels (all of which are below replacement levels) as too low and have adopted policies to raise their fertility rates; among these are France, Greece, Hungary, and Switzerland (see McIntosh, 1983, 1986).

The less developed regions present a more varied picture. The majority of the governments in less developed African countries currently view their fertility rates as being too high. However, this is a relatively recent development. In the mid-1970s, more than a third of the governments viewed their fertility rates as too high; by the end of the 1980s, this proportion had increased to nearly two-thirds. Not only did significantly more countries consider their growth rates to be too high, but greater numbers of African governments officially adopted policies to lower fertility rates during this period: The proportion of governments with policies to lower fertility increased from 25 to 49 percent. Between 1976 and 1989, the proportion of African governments providing direct support for family planning services increased from 50 to 78 percent. Still, nearly a third of the countries in Africa consider their fertility rates as satisfactory and have policies of no intervention. Three countries, Côte d'Ivoire, Equatorial Guinea, and Gabon, view their fertility levels as too low and report policies to raise them; the total fertility rates of these countries in 1980–85 were substantially above replacement levels, 7.4, 5.7, and 4.5 births per woman, respectively. The growth rates of these countries over the last several decades have been erratic and at times comparatively low. For example, the annual rate of population growth for Gabon was slightly below 0.4 percent during the 1950s and 1960s, well above 4 percent in the 1980s, and approximately 3 percent in the early 1990s.

In contrast to the dramatic changes in Africa, the fertility perceptions and policies of countries in South and East Asia have remained relatively stable over the last two decades. For instance, more than half of the governments in this region—accounting for approximately 90 percent of the population—continue to view their fertility levels as too high and have implemented policies to lower them. These countries include many of the most populous in the region: Bangladesh, China, India, Indonesia, Malaysia, Pakistan, the Philippines, and Thailand.

Ten countries in this region view their fertility as satisfactory, including Iran, Laos, Myanmar (formerly Burma), and South Korea. In addition, two countries, Cambodia and Singapore, consider their fertility to be too low and intervene to raise their rates. Many of these countries have expressed concerns over the implications of population aging, brought about by the rapid decline in fertility. This decline, accompanied by rapid economic growth and the expansion of the labor force in South Korea and Singapore, accounts in large measure for those countries' recent shifts in fertility policy.

The majority of countries in Latin America (55 percent) view their fertility rates as too high and have adopted policies aimed at lowering them: The proportion of governments adopting such policies increased from 37 percent in 1979 to 52 percent by 1989. Among the remaining countries, all except Uruguay follow policies of no intervention. Uruguay's fertility is among the lowest in the region, but is above replacement level. The government views this level as too low and has a policy to raise it. With a rapidly aging population and a high dependency ratio, the government aims not only to increase fertility, but also to reduce emigration and encourage the return of emigrants.

Despite fertility policies and perceptions, virtually all countries in Latin America provide either direct or indirect support for family planning as a human right or on health grounds. Brazil, for example, reports its fertility levels and trends to be acceptable, but promotes family planning as a component of maternal and child health programs.

Of the 12 countries of Western Asia, eight viewed fertility levels as satisfactory in 1989. These countries have policies directed at maintaining their relatively high fertility rates in order to ensure rapid growth of the indigenous population and thereby lessen dependency on foreign-born labor, promote economic growth, strengthen national security and defense, and protect national identity. Three countries, Democratic Yemen, Jordan, and Yemen, none of which are oil-exporters, but are major suppliers of labor to the Persian Gulf oil industries, view their fertility levels as too high and provide family planning services within health programs.[4] These same countries train family planning volunteers, encourage women's labor force participation, and seek to raise the educational status of women.

The only country in the region that considers its fertility too low and reports a pronatalist policy is Iraq. The rationale for this policy rests with the effort to maintain economic growth, reduce dependency on foreign labor, and strengthen national defense. The government of Iraq grants family and child allowances and paid and extended maternity leave for working women; it also limits access to modern methods of contraception.

## Accounting for trends, variations, and contradictions

Fertility, and in particular fertility regulation, has been and continues to be more problematic for governments, institutions, and organizations to address than mortality, international migration, and even population distribution.[5] Not only is it a personal and private subject, human reproduction is also fundamental to the social, economic, and political well-being of groups and of society as a whole. Governments address fertility and its regulation, especially with regard to the formulation and implementation of policy, with a greater degree of caution, care, and recognition of possible adverse political consequences than is ordinarily given to other components of population policy.

The trends over time in national fertility perceptions and policies reflect the broad changes that have occurred in the "policy climate" toward fertility and its regulation during the last two decades, as the situation in Africa illustrates. At the first African Population Conference held in Accra, Ghana, in 1971, the United Nations report of the conference (1971) stated: "Population policies in Africa must therefore not be over-preoccupied with issues such as population control which appears negative and into which much unwholesome political meanings are being used" (p. 51). Some years later in an assessment of the situation in Africa for the International Review Group of Social Science Research on Population and Development, Mabogunje and Arowolo (1978) concluded:

> In spite of the international interest in population issues, very few black African countries see population as a problem or as a major factor in the determination of their development strategy. Given the small size of the population of most of these countries (some 34 out of the 48 have a population of under 5 million), there has been a tendency to watch the annual increase with some degree of quiet satisfaction. (p. 44)

At the second African Population Conference held in Arusha, Tanzania, in 1984, the tone and conclusions of the report (United Nations, 1984) were dramatically different from those given 13 years earlier:

> Noting with great concern the rapid rate of population growth in recent years and the stresses and strains which this increasing imposes on African Governments' development efforts and on the meagre resources at their disposal, [the Conference adopts] the following Programme of Action for African Population and Self-reliant Development[:] Governments should ensure the availability and accessibility of family planning services to all couples or individuals seeking such services freely or at subsidized prices. (Annex II, pp. 2, 7)

The recent broad changes in the policy climate and the trends between 1976 and 1989 described earlier are consistent with a demographic script. Many populous or rapidly growing developing countries (for example, Algeria, China, Mexico, Haiti, India, Jordan, Kenya, and Nigeria) generally view their fertility to be too high and report policies to reduce fertility, primarily by means of providing direct support for family planning services. In addition, the countries (primarily in Europe) with relatively low fertility and some developing countries, including Iraq and Singapore, seek to raise their fertility or keep it from falling. The remaining countries have few, if any, intervention policies.

The statistics in Table 1 indicate a consistent direct relationship at the global level between current perceptions and policies and levels of fertility as measured by the total fertility rate (TFR) for 1980–85. Both the median and the mean of the TFR are: (a) lowest for countries that view their fertility as too low and wish to raise it; (b) highest for countries that consider their fertility as too high and aim to lower it; and (c) intermediate for countries that view their fertility as satisfactory or have no official position on the matter and, as to their policy stance, wish to maintain the current level of fertility or simply have no intention to intervene. Although the statistics are not time series, but cross-sectional, these results suggest that the perceptions and policies of governments toward fertility are tied to demographic trends. There is also abundant evidence that population policies are the consequences of complex interplays between levels of fertility and the social, economic, and political circumstances specific to each country (see Andorka, 1991; IUSSP, 1991; McIntosh and Finkle, 1985; United Nations, 1988a).

**TABLE 1 Median and mean total fertility rates in 147 countries, 1980–85, by government view and policy toward fertility**

| View and policy in 1989 | Median | Mean | (N) |
|---|---|---|---|
| View toward fertility | | | |
| Too low | 2.1 | 3.0 | (19) |
| Satisfactory | 3.8 | 4.0 | (50) |
| Too high | 6.0 | 5.6 | (64) |
| No official position | 3.4 | 3.8 | (14) |
| Policy toward fertility | | | |
| Raise | 2.3 | 3.2 | (19) |
| Maintain | 3.8 | 4.4 | (14) |
| Lower | 5.8 | 5.3 | (55) |
| No intervention | 4.6 | 4.4 | (59) |

An interesting illustration of this interplay is Malaysia. In a significant policy reversal, the government in 1984 discontinued its two-decades-old campaign of "two is enough" and is now encouraging families to have as many children as they can support. The government became concerned with the adverse economic consequences of rapid fertility decline, particularly with respect to its labor force, and is now calling for a population of 70 million people by the year 2100 in order to facilitate economic and social development (see United Nations, 1987a).

Another major policy shift illustrating these complexities is Singapore. Concerned with the country's rapid decline in fertility to one of the lowest levels in Asia (1.7 children per woman from 1980–85), and the implications of this trend for the labor force and related economic and social considerations, the government shifted its policy in the early 1980s from promoting two children per family to promoting a three-child family. The Singapore government granted tax rebates to families with three children, as well as priorities to such families for primary school registration and public housing allocation, and it extended paid leave to mothers to care for their infants (see Cheung, 1989; Saw, 1986 and 1989).

The case of Nigeria, the most populous country in sub-Saharan Africa, also illustrates the complex interactions between demographic pressures and social, economic, and political considerations. Prior to 1985 the Nigerian government followed a policy of nonintervention in population matters. Although acknowledging the country's high rate of population growth, the government did not feel that the growth rate constituted a serious obstacle to social and economic development. The official viewpoint was that higher rates of economic growth and progress would reduce the birth rate in the long run. As information from censuses, surveys, and studies became available and an economic downturn occurred with a sharp decline in oil prices in 1983, a growing awareness of dwindling resources and a perceived lack of progress in many social and economic areas resulted in the initiation of a new policy aimed at lowering population growth and fertility. The policy was officially adopted in 1988 (United Nations, 1988b).

As these examples make clear, governments are capable of changing population policies rapidly and dramatically. Understanding policy evolution requires the study of demographic, social, economic, and political factors and circumstances of countries.

A related issue is the discrepancies between fertility policies and their implementation by governments. Such discrepancies may be genuine reflections of the existing state of affairs within a country. For example, there may be a general administrative inability to implement desired programs efficiently and effectively. Sound and comprehensive fertility policies may have been formulated, but shortages of human and financial resources as well as poor management may hamper the government's attempts to implement programs that its policies promote.

A government's lack of commitment or resolve may also cause its programs to falter. For example, a government may formulate a population policy in response to external pressure from bilateral and multilateral organizations (sometimes referred to as a "conditionality"); however, once aid is in hand, implementation of the policy flags.

A factor that contributes to the weakening of government commitments to population programs is the ideological differences among various parties, organizations, and interest groups on such issues as induced abortion, contraception, and sterilization. The establishment of fertility regulation programs and services may be viewed by an ethnic, racial, or religious group as an organized attack by the government on the group's strength, influence, status, values, and beliefs. To avoid confrontation with these groups, the government may scale down, postpone, or even abandon proposed fertility regulation programs.

Governments are not monolithic actors, but are composed of numerous groups and persons whose perceptions and objectives are shaped in part by their specific areas of responsibility. With regard to fertility limitation, the views and goals of those working in a health ministry are likely to differ markedly from those in the planning or defense ministries. Therefore, the source of the views that are represented in official government statements, reports, and plans and the extent of the diversity of opinion within a government are both factors of importance when gaps and contradictions between policy statements and the implementation of programs are examined.

## Conclusions, reflections, and speculations

Significant shifts have occurred in government perceptions and policies toward fertility since the mid-1970s. In the less developed countries, a shift has taken place from the "satisfactory" fertility perception and no intervention policies to the "too high" fertility perception and policies to lower fertility. In addition, the region where this shift was found to be the most dramatic was Africa, where the proportion of countries with policies to reduce fertility increased from 25 to 49 percent between 1976 and 1989. Among more developed countries, modest increases were also observed in the proportions of countries that view their fertility as too low and that have policies to raise it.

Increasing numbers of countries are also reporting policies of direct support for the provision of contraceptive methods. Currently, nearly three out of four countries report policies of direct support for family planning services. Again, this shift is notable in the case of Africa, where the proportion of countries providing direct support for the provision of birth control methods increased from 50 to 78 percent between 1976 and 1989.

Among less developed countries the shifts in the fertility perceptions and policies of governments are consistent with widespread changes that have occurred in the policy climate toward fertility and its regulation. For example, in contrast to the rhetoric of intergovernmental conferences in the early 1970s, in which fertility intervention policies and family planning programs were frequently viewed with suspicion and linked to excessive Western influence, the declarations and recommendations of conferences held in the early 1980s, such as the Regional African Conference in Arusha, Tanzania, the Regional Conference in the Arab World in Amman, Jordan, and the International Population Conference in Mexico City, reflect recognition of the consequences of sustained high fertility and are largely supportive of policies and efforts to provide family planning services.

Current perceptions and policies of governments toward fertility are often found to be consistent with actual fertility levels. Generally, countries with policies designed to lower fertility had the highest fertility; those with policies aimed at raising it had the lowest; and countries with policies devised to maintain fertility or not intervene had intermediate fertility levels. Government policies toward fertility appear, therefore, to be tied to demographic trends, although they may also develop from social, economic, and political circumstances.

A plausible explanation for the observed discrepancies between fertility policies and their implementation is a government's inability or lack of motivation to push fertility regulation programs through. The discrepancies between what is written in the capital and what is done in the field appear to reflect the state of administrative, social, and political affairs existing within a government and country.

Important changes in population and fertility policy have taken place in the four decades from 1950 to 1990. With the exception of a few countries, including Egypt and India, the 1950s were not noted for government intervention to limit fertility or population growth. Most governments, and many nongovernmental agencies providing development assistance, considered fertility regulation to be outside their purview.

In the 1960s, Western governments became concerned with high rates of population growth and fertility. During this decade the average annual rate of global population growth peaked at 2.1 percent. The United States government's earlier policy of no foreign aid for population assistance was reversed in the 1960s. The Johnson administration launched major initiatives to provide family planning methods and to reduce high rates of population growth in less developed countries.

The 1970s was a period of controversy. Led largely by the United States and a few other Western industrialized countries concerned with the consequences of rapid population growth and high fertility, the stormy 1974 World Population Conference adopted the World Population Plan of Action (WPPA) (see Finkle and Crane, 1975, and Berelson, 1975, for details of the WPPA and conference). Despite the consensus achieved at the conference and the adoption of the WPPA, the majority of developing countries did not share the concern of Western countries and continued to question attempts to redirect development assistance from larger social and economic issues to programs of fertility control.

By the time of the International Population Conference in 1984, the situation had changed dramatically. The objections of the less developed countries, clearly evident at the 1974 conference, were strikingly absent during the 1980s (Finkle and Crane, 1985). Increasing support for population policies and programs came from international and nongovernmental organizations, especially donor institutions, and the climate of opinion toward population policies and programs aimed at influencing fertility and population growth had become more tolerant. During the 1980s the governments of less developed countries increasingly formulated and adopted policies to reduce their rates of population growth through the reduction of fertility levels.

What does the future hold? The demographic outlook for the end of the twentieth century is a world population of nearly 6.5 billion, with most of the increase occurring in poor, less developed countries, and the rapid growth of Africa's sub-Saharan population. Because governments are capable of changing population policies, especially those relating to human reproduction, rapidly and substantially in response to new social, economic, and political conditions, the policies adopted by the governments of many less developed countries in the 1970s and 1980s aimed at reducing fertility should not be viewed as fixed. New governments may choose to distance themselves from the policies of previous administrations. There may also be resistance to policies as a result of general disenchantment with a government's lack of progress in addressing its most pressing problems. A number of countries, particularly in Africa, are facing the task of implementing their fertility policies. Most of these countries face substantial, and in some cases overwhelming, obstacles, such as inadequate financial resources, social unrest, slow or stagnant economic growth, and political instability. As a result, the prospects for implementing their fertility policies are uncertain.

Finally, low fertility levels and high labor immigration remain central concerns for many of the more developed countries (primarily in Europe and Japan), as well as some comparatively wealthy less developed countries. As the demand for labor exceeds the population capacities of these countries (or is anticipated to exceed them), the governments are under increasing pressure to take action. Advocacy of higher fertility levels in this instance may be politically palatable, but its demographic impact is likely to be insignificant or too slow in coming to satisfy a country's current or short-term labor needs. Although the immigration of foreign labor may occur almost instantly, its promotion as a course of action is problematic. In many societies, foreign workers and their families are perceived as being significantly different in terms of race, religion, language, and ethnicity. Such perceived differences contribute to opposition and resistance to their immigration, hostility and, in some cases, violence toward the migrants, and demands for the repatriation of foreign workers and their dependents.

## Notes

1 The Eisenhower quotation is cited in *Population: A Clash of Prophets*, edited by Edward Pohlman (New York: Mentor Books, 1973, p. 485). In February 1968, former President Eisenhower expressed a change of view: "Once, as President, I thought and said that birth control was not the business of our Federal Government. The facts have changed my mind. I have come to believe that the population explosion is the world's most critical problem. Failure would limit the expectations of future generations to abject poverty and suffering, and bring down upon us history's condemnation."

2 Additional information about the United Nations population policy data base may be found in United Nations, 1990b.

3 At the time of assessment in 1989, the government of Romania had a policy limiting the provision of contraceptives. As noted previously, this policy was changed when the Ceauşescu government was overthrown. Currently Romania provides direct support for public access to contraceptives.

4 On 22 May 1990, Democratic Yemen and Yemen merged to form a single country, Yemen. The information presented in this article describes the situation prior to the merger.

5 In the past, even nongovernmental organizations had difficulty dealing with fertility regulation matters. For example, with regard to determining the agenda of a conference for the International Union for the Scientific Study of Population, consider the following quotation taken from a letter written by Raymond Pearl to Margaret Sanger, dated 19 April 1926:

> The whole project of organizing an international union of population, having as adhering bodies the leading scientific organizations of the world interested in this problem, demands, as I see it, that birth control, or Neo-Malthusianism, shall not appear as being the dominant element in the organization or plan. (Pearl, 1926)

## References


Andorka, Rudolf, and Raúl Urzúa. 1991. *The Utilization of Demographic Knowledge in Policy Formulation and Planning*. Liège: International Union for the Scientific Study of Population.

Berelson, Bernard. 1975. "The World Population Plan of Action: Where now?," *Population and Development Review* 1, 1: 115–146.

Chamie, Joseph. 1988. "Les positions et politiques gouvernementales en matière de fécondité et de planification familiale," in *Population et sociétés en Afrique au sud du sahara*, ed. Dominique Tabutin. Paris: L'Harmattan, pp. 167–190.

Cheung, Paull P. 1989. "Beyond demographic transition: Industrialization and population change in Singapore," *Asia-Pacific Population Journal* 4, 1: 35–48.

Finkle, Jason L., and Barbara B. Crane. 1975. "The politics of Bucharest: Population, development, and the new international economic order," *Population and Development Review* 1, 1: 87–114.

———, and Barbara B. Crane. 1985. "Ideology and politics at Mexico City: The United States at the 1984 International Conference on Population," *Population and Development Review* 11, 1: 1–28.

Hardee-Cleaveland, Karen, and Judith Banister. 1988. "Fertility policy and implementation in China," *Population and Development Review* 14, 2: 245–286.

International Union for the Scientific Study of Population (IUSSP). 1991. *Population Policy in Sub-Saharan Africa: Drawing on International Experience*. Liège: IUSSP.

Mabogunje, A. L., and O. Arowolo. 1978. "Social science research on population and development in Africa south of the Sahara," *International Review Group of Social Science Research on Population and Development*, Appendix 7. Mexico City.

McIntosh, C. Alison. 1983. *Population Policy in Western Europe: Responses to Low Fertility in France, Sweden and West Germany*. Armonk, NY: M. E. Sharpe.

———. 1986. "Recent pronatalist policies in Western Europe," *Population and Development Review* (Supp. 12): 318–334.

———, and Jason Finkle. 1985. "Demographic rationalism and political systems," in *IUSSP International Population Conference*, Florence, Italy, Vol. 3, pp. 319–329.

Pearl, Raymond. 1926. Letter to Margaret Sanger, 19 April 1926, in "The Papers of Margaret Sanger." Washington, DC: Library of Congress.

Saw, S. H. 1986. "A decade of fertility below replacement level in Singapore," *Journal of Biosocial Science* 18, 4: 395–401.

———. 1989. "Muslim fertility transition: The case of Singapore Malays," *Asia-Pacific Population Journal* 4, no. 3: 31–40.

United Nations. 1974. *Report of the United Nations World Population Conference, Bucharest, 19–30 August 1974*. New York: United Nations.

———. 1979. *World Population Trends and Policies: 1977 Monitoring Report*. New York: United Nations.

———. 1980. *World Population Trends and Policies: 1979 Monitoring Report*. New York: United Nations.

———. 1982. *World Population Trends and Policies: 1981 Monitoring Report*. New York: United Nations.

———. 1985. *World Population Trends, Population and Development Interrelations and Population Policies: 1983 Monitoring Report*. New York: United Nations.

———. 1987a. *Case Studies in Population Policy: Malaysia*. New York: United Nations.

———. 1987b. *World Population Policies: Afghanistan to France*. New York: United Nations.

———. 1988a. *World Population Trends and Policies: 1987 Monitoring Report*. New York: United Nations.

———. 1988b. *Case Studies in Population Policy: Nigeria*. New York: United Nations.

———. 1989a. *Trends in Population Policy*. New York: United Nations.

———. 1989b. *World Population Policies: Gabon to Norway*. New York: United Nations.

———. 1990a. *World Population Trends and Policies: 1989 Monitoring Report.* New York: United Nations.

———. 1990b. *International Transmission of Population Policy Experience.* New York: United Nations.

———. 1990c. *Global Population Policy Data Base 1989.* New York:United Nations.

———. 1990d. *Results of the Sixth Population Inquiry among Governments.* New York: United Nations.

———. 1990e. *World Population Policies: Oman to Zimbabwe.* New York: United Nations.

United Nations, Economic Commission for Africa. 1971. *Report of the African Population Conference, Accra, 9–18 December 1971.* Addis-Ababa: United Nations.

———. 1984. *Report of the Second African Population Conference, Arusha, 9–13 January 1984.* Addis-Ababa: United Nations.

Zeng, Y. 1989. "A policy in transition," *People* 16, 1: 20–22.

# Political Leadership and Policy Characteristics in Population Policy Reform

JOHN W. THOMAS
MERILEE S. GRINDLE

NATIONAL POLICIES AND PROGRAMS intended to influence population size and growth rates have become commonplace in developing countries. Yet their prevalence is not synonymous with success. In many countries, reproductive behavior has been slow to change and growth rates remain high. This reflects the degree to which population policies, intended to alter reproductive behavior, are controversial, politically charged, and often ineffectively implemented.

This chapter addresses the issue of reform in population policy. By population policy reform, we refer to any deliberate effort on the part of government to redress perceived errors in prior and existing policies to limit or otherwise affect reproductive behavior.[1] We take instances of efforts to alter existing population policy (including efforts to introduce explicit policy where none existed) as a unit of analysis that can be assessed to respond to several questions:

How do issues of reform come to the attention of government decision-makers?

How do such issues become part of an agenda for governmental action?[2]

What criteria do policy elites apply when they consider recommendations for reform?[3]

What factors influence how reforms are pursued and sustained after they have been introduced?

In this chapter we are particularly interested in the role of policy elites—those officially responsible for making authoritative decisions for government—in reform initiatives. We believe that much of the explanation of policy change rests on a more systematic understanding of the role of these public officials. We write as analysts who are concerned primarily with the process and dynamics of policy change rather than as experts with long experience in the population field.

In what follows, we first present a summary of an analytic framework that we have developed on the basis of a series of economic policy reform initiatives,[4] and then assess it in light of 16 cases of population policy change that we have reviewed—considering the experiences of China, the Dominican Republic, Egypt, Fiji, India, Indonesia, Kenya, Lebanon, Malaysia, Mauritius, Mexico, Pakistan, the Philippines, Singapore, Thailand, and Zambia (see Table 1). We draw on these cases for examples, but do not develop any of them fully; for greater detail, readers should consult the references in Table 1. If our framework is useful in understanding the dynamics of the 16 cases of population policy reform, then it may have some practical worth for those concerned about the introduction and pursuit of change in this policy sector.

## Policy elites and policy change: A framework for analysis

In *Public Choices and Policy Change* (1991), we identified a general process of policy and institutional reform that is determined by context, circumstance, and policy characteristics. The framework we developed indicates that contextual conditions, agenda-setting circumstances, and policy characteristics influence the perceptions and concerns of policy elites and shape the nature and scope of conflict surrounding efforts to introduce change. The case studies we analyzed in that earlier work suggest that assessing context, circumstance, and policy characteristics can account for much variability in the outcome of reform initiatives across countries, issues, and time.

According to our framework, an initiative to reform population policy emerges within a pre-existing situation composed of (1) the orientation of policy elites in terms of their values, expertise, experiences, and loyalties, and (2) the historical, international, political, economic, and administrative characteristics of a given country, as well as the values of societal groups and the cultural context. These factors form the context within which decisions are made. Those involved in making decisions about policy are constrained by such factors, which color their perceptions of reality and set limits on what is possible within a given society. However, this contextual situation not only constrains policy options, but also provides scope and opportunity for the influence of ideas and values, for the creation of coalitions of support, and for the pursuit of change.

Our framework also indicates that the circumstances surrounding specific issues are important. The dynamics of issues that appear on reform agendas when policy elites perceive that a crisis exists are distinctly different from the dynamics that surround initiatives when decisionmakers believe that no crisis exists. We distinguish, therefore, between circumstances of perceived crisis and politics-as-usual.[5] Given different circumstances, decisionmakers will be dominated primarily by concerns about macropolitical stability and legitimacy or about micropolitical and bureaucratic costs and benefits. Choices among available options vary in part because policy elites apply different criteria to the decisions they make within differing situations. The options considered and concerns voiced clearly reflect contextual factors such as ideological preferences, political organization, and historical experience. How particular circumstances are perceived by policy elites thus serves as a bridge between the "embedded orientations" of individuals and societies and the kinds of changes considered by decisionmakers confronted with specific policy choices.[6]

In turn, characteristics of particular policies form a bridge between what is decided and what consequences, in terms of social and bureaucratic conflicts, are likely to follow. Policy characteristics determine who experiences the costs and benefits of altered policies and under what conditions and timing they are likely to react to the consequences of state action. Some costs and benefits are distributed in a way that encourages such public response as the mobilization of support or opposition groups. In other circumstances, bureaucratic response and resistance are likely to emerge in the wake of proposed changes. These arenas of conflict—public and bureaucratic—influence the stakes for political regimes, policy elites, and reform outcomes in any effort to introduce change. The framework indicates that, depending on the area of conflict and level of the stakes involved, different kinds of resources are needed to overcome resistance or lessen the risks of introducing change. The consideration of what these resources are and whether they are available to reformers leads us to reassess the broader context of values, institutions, and experiences with which we began.

The framework presented in *Public Choices and Policy Change* does not predict what choices will be made or how successfully particular reforms will resolve particular public problems. Instead, it systematizes thinking about how context influences particular situations, how circumstances shape options, how options are sorted out in terms of their political, technical, bureaucratic, and international implications, and how policy characteristics affect conflict and the resources needed to manage it in the introduction of reform. Mapping out these relationships is an aid to strategic thinking about introducing and sustaining reform.

## Population control: The dynamics of policy reform

In most cases, the role of government officials in shaping the timing and content of population policy change is noticeable. Rarely, in fact, does population policy become an agenda item through social mobilization and pressure, although the involvement of domestic and international organizations is almost universal. Rather, decisionmakers' perspectives about the rela-

**TABLE 1 Population policy case histories**

| Country | Date | Description | Source |
|---|---|---|---|
| China | 1970 | The Chinese government initiated a "later-longer-fewer" family planning program to contain population growth. | Hardee-Cleaveland and Banister (1988); Whyte and Gu (1987); White (this volume) |
| | 1979 | China implemented a one-child-per-family population control policy. | |
| Dominican Republic | 1968 | President Balaguer created the National Council on Population and the Family to coordinate implementation of an integrated family planning and reproductive health program, ending the government's laissez-faire attitude toward population growth. | Warwick (1982) |
| Egypt | 1966 | President Nasser issued an edict instituting a Supreme Council for Family Planning that added family planning to the services all health clinics should provide. The government's policy shifted from an exclusive resource-development focus to include population control. | Warwick (1982) |
| Fiji | 1962 | The government of Fiji officially adopted family planning as part of its health services, created a Family Planning Association in 1963, and reorganized its Medical Department to expand fertility control programs. | Hull and Hull (1973) |
| India | 1966 | Indira Gandhi shifted India's family planning policy from one of strict voluntarism to a rigorous, high-priority effort with ambitious fertility control programs. | Finkle (1972); Maru, Murthy, and Satia (1986); Warwick (1982) |
| | 1977 | Outraged at government abuses in family planning, Indians elected the Janata government, which returned to a voluntary population control program. | |
| Indonesia | 1975 | President Suharto made family planning a priority and delegated responsibility to regional governments to slow population growth. | Pyle (1985); Snodgrass (1979); Warwick (1986) |

**TABLE 1 (continued)**

| Country | Date | Description | Source |
|---|---|---|---|
| Kenya | 1967 | The Ministry of Health launched a Family Planning Program to provide information and services in all government hospitals and health centers. Domestic opposition and lack of support from top officials, including President Kenyatta, meant only nominal implementation of the program. | Warwick (1982); Radel (1973); Frank and McNicoll (1987) |
| Lebanon | 1969 | Lebanon's de facto, but never articulated, policy in support of family planning became apparent when, in spite of laws prohibiting birth control, shipments of contraceptives were distributed to clinics. | Warwick (1982) |
| Malaysia | 1966 | The Malaysian government formed the National Family Planning Board, changing its position with regard to family planning from hostility to support. | Ahmad and Chin (1975); Lee, Ong, and Smith (1973); Ness and Ando (1971) |
| Mauritius | 1970 | The Ministry of Health took over an independent family planning association, forming a Family Planning Division to implement a comprehensive program. | Greig (1973) |
| Mexico | 1972 | In an abrupt policy reversal, President Echeverría established a family planning program within maternal and child health-care centers, later expanding to hospitals and public and private organizations under the coordination of a Council of Maternal and Child Health. | Alba and Potter (1986); Warwick (1982) |
| Pakistan | 1959 | President Ayub declared the initiation of a drive for population control through contraceptive education and family planning. | Finkle (1972) |
| Philippines | 1969 | President Marcos created a Population Commission to institute a national population control effort in addition to existing regional and private family planning units. | Lopez and Nemenzo (1976); Ness and Ando (1971) |
| | 1972 | An aggressive policy, including sterilization, was effected. | |

**TABLE 1 (continued)**

| Country | Date | Description | Source |
|---|---|---|---|
| Singapore | 1959 | The Legislative Assembly voted to fund the Family Planning Association for voluntary population control. | Thompson and Smith (1973) |
| | 1966 | The government assumed the responsibilities of the Family Planning Association and intensified efforts to distribute contraceptives. | |
| Thailand | 1964 | The Ministry of Health initiated an experimental family planning service. | David and Viravaidya (1986); Knodel, Havanon, and Pramualratana (1984); Weeden et al. (1986) |
| | 1970 | The government promulgated a population policy, establishing a national family planning program and founding the Planned Parenthood Association of Thailand. | |
| Zambia | 1979 | The government announced a family planning policy in a reversal of a previously pronatalist position. | Hopkins and Siamwiza (1989) |

tionship between population growth and economic growth and about maintaining social stability significantly affect how issues of population growth and limitation are added to government agendas. Much of the overt political drama of population policies occurs in their implementation, when specific initiatives are embroiled in bureaucratic interactions and public responses to governmental efforts to alter individual and household-level behavior. How, then, do context, circumstance, and policy characteristics influence this process of agenda setting and decisionmaking in population limitation policies?

## The context of policy choice

We hypothesize that specific policy choices are the result of activities that take place largely within government and that are shaped by policy elites who bring their own perceptions, commitments, and resources to bear on the content of reform initiatives, but who are also influenced by the actual or perceived power of social groups that have a stake in reform. Our framework for understanding the policy process begins with two sets of contextual factors. One set focuses on background characteristics of policy elites; another emphasizes the constraints and opportunities created by the broader contexts within which they seek to accomplish their goals. The framework focuses first on a series of factors that set the stage for particular reform initiatives.

The framework captures important features of the country-specific case histories of population policy that we reviewed. Most of these case histories reveal that the chief executive is the principal actor in changes in population policy. This person's attitudes, values, positional resources, and training are frequently central to the introduction of—or the failure to introduce—population reform initiatives. Ministers of health and planning, as well as their technical corps, also often figure as critical filters through which population initiatives flow. These officials, as well as domestic and international interest groups, in their efforts to bring about change or forestall it, attempt to alter the perceptions of the chief executive whose support is deemed essential to introducing new or adjusting existing policy. The case histories of China, Egypt, India, Indonesia, Kenya, Mexico, Pakistan, and the Philippines all reflect the importance of the pre-existing characteristics of policy elites in the introduction of reform. The factors that appear most important in accounting for policy change are the ideological predispositions, the expertise and training, the positional resources, and personal attributes and goals of the decisionmakers.

In Mexico in 1972, for example, President Luís Echeverría was the central figure in the abrupt reversal of decades of pronatalist policy in the country. His powerful position at the apex of the Mexican political system, his concerns about the destabilizing effect of rapid population growth, and his concerns about the future of his country's economic development were important factors in his public announcement that Mexico needed to address its population problem. In Kenya in the 1960s, President Jomo Kenyatta's desire to maintain his position as leader and symbolic unifier of the country made him keep his distance from the potentially controversial population debate. Kenyatta, an anthropologist by training, and acutely sensitive to cultural norms, never lent support to the national population program. His concerns about his position, about tribal relationships, and about colonialism and the importance of national sovereignty made him view population growth positively and made him unwilling to encourage international donors and domestic groups advocating population control.

In the post-independence period in Egypt, India, the Philippines, and Zambia, Nasser, Nehru, Marcos, and Kaunda all believed that economic development would resolve the population problem, and they kept the issue off their governments' agendas for a number of years, despite pressure from population control groups. When these same leaders became convinced that population growth acted as a constraint on economic growth, serious government attention to the issue became a reality. This was also the case in China, where national leaders became convinced that population growth inhibited their country's development.

Similarly, few cases of population policy reform are presented here without discussions of the broader contextual factors that help shape the perceptions of decisionmakers and help set the boundaries of policy choice.[7] A country's religious identities and cultural norms regularly limit policymakers' options. In Lebanon throughout the 1960s and 1970s, for example, sectarian conflict severely constrained open discussion of population policy; and in Singapore, urban development and prosperity seem to have played an important role in the adoption and rapid extension of population control measures. In the Dominican Republic, Kenya, Mexico, and the Philippines, concern about the response of the Catholic Church helped keep population policy off government agendas for considerable periods.

Interest groups have imposed constraints on options for population policy in a number of countries. In general, those who mobilize to promote the introduction of population control policies are concerned about issues related to population growth and national development or to health. Evident among these groups are foreign donors and expatriate advisors. They are often joined in pressuring government by small groups of intellectuals, academics, doctors, and voluntary-agency leaders—elite groups lobbying for mass policy. Organizations with a mass base, such as the church or ethnic and regional groups, are often opposed to population control policies. These organizations tend to base their positions on cultural, religious, or ethnic values, and they appeal less to the policymakers than to their own followings.

In many of the cases reviewed here, however, surprisingly little public discussion and debate surrounded the initiatives to introduce population control policies. In one case, that of Malaysia in the mid-1960s, analysts even commented on "the absence of politics" in the introduction of population policy. Centralized party control and decisionmaking kept the issue out of public forums. In Mexico, President Echeverría's announcement of a population policy in 1972 took most observers by surprise. The dearth of debate reflects, in part, the extent to which decisionmaking is a closed and executive-centered process in many developing countries. However, decisionmakers were aware of the constraints set by societal groups and interests. Indeed, they knew well the potential risks of introducing policies that offended religious and ethnic organizations or that conflicted with well-established cultural values. Such awareness often determined whether policy change should or could be introduced.

In the Dominican Republic, private family planning efforts were given tacit approval for many years by a government that was unwilling to challenge the Catholic Church publicly through an official policy. In Kenya, despite strong support for a population control policy within the Ministry of Economic Planning and Development, boundaries set by ethnic rivalries and the fear of political instability limited the government's commitment to an explicit policy. In the Philippines in the 1960s, international and domestic agencies' extensive mobilization and advocacy of family planning were countered by leaders' concern about potential opposition from the Catholic Church. The implicit threat of offending religious groups set boundaries on the speed of family planning policy change in the 1960s and 1970s in Indonesia, Lebanon, and Malaysia. Countries rarely experienced public demand for population policies, so governmental leaders frequently were able to keep the population issue off their policy agendas for considerable periods of time, despite pressures from donors, nongovernmental organizations, and their advocates.

### The circumstances of setting agendas and making decisions

The general context of the policymaking process in developing countries forms a backdrop for the circumstances that place particular issues on the agenda for government decisionmaking. As we show in *Public Choices and Policy Change*, many reform issues emerge and are considered when policy elites believe that a crisis exists, that they must do something about a situation or face grave consequences in the short term. But reform issues also emerge under conditions that can be described as politics-as-usual, in which change is considered desirable but the consequences of inaction are not considered immediately threatening to the decisionmakers or to the regime. Circumstances of crisis or of politics-as-usual alter the dynamics of decisionmaking by raising or lowering political stakes for policy elites, altering the identity and hierarchical level of decisionmakers, and influencing the timing of reform.

Almost all of the cases of population policy initiatives reviewed here conform to the politics-as-usual circumstance. Advocates of population policy tried to convince decisionmakers in these cases that population growth constituted a crisis situation for the country's development. Detailed accounts of how the issue was handled belie this sense of urgency.

Population growth is usually a chosen problem, selected by decisionmakers as a priority when they become convinced that public initiative is required.[8] They have considerable latitude for taking up or ignoring such problems, and while, in many of the cases we reviewed, groups mobilized to press for population control measures, there is little to suggest that decisionmakers were forced to consider the issue. In Egypt, Indonesia, Kenya, Malaysia, Mauritius, Mexico, and the Philippines, concern about population growth and the need for family planning services has existed for decades, as have campaigns to convince officials to take action. Foreign donors were often in evidence with advice and promises of funding, yet policy elites took up the issue according to their own sense of timing and priority.[9]

In the Philippines, the 1960s was a decade of strong pressure for population regulation policy from domestic organizations and the United States Agency for International Development. Not until the elections of 1969 had ensured his tenure in office, however, did President Marcos take any substantive action on population policy. In the other countries just cited, policy elites made their own decisions about whether and how to address the issue of population policy. Few such officials were motivated by an immediate sense that their response to this issue could make or break their political careers.

Our earlier study indicated that politics-as-usual reforms tend to involve middle-level decisionmakers and managers. In population policy, this has often been the case. Population policy has frequently been the priority of technical groups within ministries of planning or health, as was the case in Kenya and Malaysia. When chief executives played pivotal roles in advocating population policy, they were essentially giving other government officials the go-ahead. Elsewhere, chief executives have been the principal actors in keeping population control off government agendas.

Incremental changes in policy tend to dominate politics-as-usual agenda-setting circumstances. Many of the initial decisions of governments in population policymaking were made to give public sanction to activities that were already under way within the public sector or to introduce official activities that replicate private initiatives, such as the provision of family planning services through clinics and hospitals. Even in Mexico, President Echeverría's dramatic public reversal on population control tended to obscure the fact that government hospitals had well-established contraceptive services. In the Dominican Republic, Fiji, Mauritius, and the Philippines, public organizations took over or replicated long-existing private initiatives. There are cases of agenda-setting circumstances, for example those in Kenya and Mexico, in which startling and innovative reversals of stated policy have taken place, suggesting that the incremental nature of politics-as-usual agenda-setting is not always a given. Once policies have been put in place, however, most subsequent efforts to reform population measures come about gradually.

In the 16 cases we reviewed, the timing of decisions about population control was rarely urgent; policy elites appeared to have considerable leeway in introducing change when the time seemed propitious and in delaying action when it did not. Periods of several years passed in many cases between a governmental decision to do something about population growth and the introduction of substantive measures to accomplish this goal. The Dominican Republic, Indonesia, Malaysia, Mauritius, and Singapore are countries in which population policy was under consideration for long periods before any decision was reached or any action taken.

In these politics-as-usual circumstances, policy elites played important roles in selecting the moment for reform, shaping the terms of debate, and generating consensus about the need for policy change. Their views about how to address the problem of population—through economic development or through population control policy—generally determined whether or not the issue was placed on the agenda.

If such officials retain some autonomy to shape policy, what factors influence their decisions? Decisionmakers tend to consider options according to their understanding of (1) the technical aspects of the issues; (2) how changes will affect micropolitical relationships within the bureaucracy and between groups in government and their clienteles; and (3) what impact reform measures will have on the regime's overall stability and support. Decisionmakers consider as well the roles of pressure, support, or opposition from international aid agencies and foreign governments. But these criteria of judgment shift in importance depending on the circumstances surrounding an issue. Thus, crisis-ridden reforms are likely to be assessed first in terms of their expected macropolitical impact: Will such reforms cause major outbursts of political opposition, lead to the threat of a coup, or contribute to electoral losses? In less fraught circumstances, reforms are considered primarily in terms of how they will affect relationships within the bureaucracy or within narrow clienteles: Will they affect, for example, the ministerial budgets in important ways, encourage noncompliance among bureaucrats, or generate disaffection among specific interest groups? We found that different processes of agenda-setting result in the adoption of different decision criteria by decisionmakers.

In all of the cases of population policy that we reviewed, policy elites were influenced by technical analyses of the population problem. In particular, projections of population and economic growth rates figured centrally in convincing a number of national leaders to act on population policy. This was a principal means used by proponents of reform to raise the specter of crisis for policy elites. In many cases—Indonesia, Kenya, Malaysia, Mauritius, Mexico, Singapore—official reports, commissioned studies, and national and international conferences influenced the perspectives of decisionmakers. Macropolitical factors having to do with religious and ethnic support and opposition, the potential of cultural and religious conflict to destabilize the political system, and concerns about the political implications of projected unemployment appear to have been of central importance in shaping the decisions made by policy elites. Along with these macropolitical concerns, relationships with international donors and concern over the future of such connections were critical for a number of decisionmakers. Micropolitical and bureaucratic concerns were less important to them, in contradiction to our expectations. The differences between these 16 cases and our earlier work may be explained in part by the perceptual links between population policy and economic growth and between fertility control and religious and cultural values in a society. These factors appear to have great impact on choices about how to respond to the population problem in specific countries.

### Policy characteristics

Contextual factors loom large in the process of agenda-setting and decisionmaking when reform initiatives emerge and are considered. In contrast, the characteristics of particular policies appear to be the critical factors determining the outcome of implementation efforts. Predicting the reaction to a policy change or new program, or where opposition or problems are likely to occur, is a central consideration for those advocating a particular policy. Our earlier research on policy change shows that reforms characterized by broadly dispersed costs and by effects that are quickly visible are likely to elicit strong and immediate public reactions, in which the stakes for government are high. Other reforms, characterized by costs concentrated in government or a limited public and by effects that are not immediately visible, tend to be carried out or subverted largely within the bureaucratic arena, where the potential for failure is great but the stakes for government are relatively low.

The ability to predict reactions to policy changes and new programs is especially important for makers of population policy. While formal population programs exist in virtually every country, in most programs there are major gaps between policy statements and what is implemented. In many cases, these gaps result from the failure of policy elites and managers to consider the nature of conflict that occurs during the implementation of their programs.

The specific cases studied show that initiatives to limit population growth engender reactions within the bureaucracy and between the bureaucratic administrators and the potential users of contraceptive technology. Noncompliance with population policies within the bureaucracy results from resistance to change, loss of power, or failure to accept new responsibilities, and may range from overt opposition to quiet sabotage or inaction. In some cases, alliances between factions or individuals in the bureaucracy and external interest groups or clienteles can result in piecemeal sabotage of the intent of the policymakers (see Grindle, 1980).

Bureaucratic capacity was another important problem in many of the case studies of population policy that we reviewed. In many countries, new cadres of family planning administrators had to be recruited and trained. In others, rural health-clinic personnel had to be familiarized with contraceptive techniques and health issues. Developing this type of capacity emerged as a critical constraint on implementation in many cases.

If policymakers focus on those characteristics of a policy that determine its costs and benefits and on those that influence how quickly the policy's impact is visible, they can estimate the nature of the response to the new policy. In our framework, costs and benefits include political, personal, and organizational resources, as well as economic resources.

*Concentration of costs in government* In population policy, public agencies and officials must take on new responsibilities, adopt new forms of behavior, and face the potentially hostile reaction of the people they are expected to serve. Resistance or opposition to such change is likely. The official response to policy change in Kenya illustrates this situation. Although some political leaders gave public support to the family planning program in the late 1960s and early 1970s, the bureaucracy, sensing limited top-level political support and considerable public opposition, put forth only a limited effort to implement the program. In Egypt, where political support for a family planning program was strong, the administrative officials charged with implementation did not comply with the political mandate.

*Dispersion of benefits* Reforms that concentrate costs in the government often have broadly dispersed benefits that become visible only in the long term. The public may benefit eventually from the reduction of population growth, and individual citizens may benefit more quickly from the availability of new services, but these medium- and longer-term benefits do not tend to mobilize active support. The direct impact of population programs is initially borne by officials and institutions that are required to alter traditional behavior. The public support such reforms may generate could eventually counteract the opposition to change that may arise in the bureaucracy, but administrators are likely to become aware of the costs of reform long before the public appreciates its benefits.

*High administrative and technical content* If the administrative content of a policy is high or it is technically complex, that policy will require the coordinated efforts of public officials and institutions to ensure that it is carried out. Population programs clearly require both extensive administration and the delivery of services that are often technically complex. Such programs in Indonesia and Mexico, for example, required time to develop new administrative and technical capacities. In such cases, the public will not be affected immediately, and implementation of the program depends on competence and support in the bureaucracy.

*Public participation* Population programs have high administrative and technical costs and, at the same time, require extensive public participation, a characteristic that is less true of the macroeconomic reforms we assessed in our earlier work. In order for a family planning program to be effective, large numbers of individuals will have to alter their beliefs, values, and behavior.

*Long duration* The longer the time needed to implement a reform program, the less likely it is that conflict and resistance will emerge in a public arena and the more likely that administrative compliance and capacity within the system will determine the ability to implement and sustain the reform. Most population programs in our case studies were formally enacted or announced in the 1960s and, decades later, are still evolving. For example, Indonesia's family planning board was established in 1968 and the National Family Planning Coordinating Board was initiated by presidential decree in 1970. The program was administered successfully throughout the 1970s and underwent major modifications in the 1980s.

Because initiatives in population policy regularly display most of the characteristics described above, implementation efforts tend to generate conflict in bureaucratic arenas. Such policies begin as bureaucratic arena programs because they carry dispersed and long-term benefits and the costs are concentrated in government. In some cases, however, public conflict does ensue, usually as the costs of specific parts of a program become visible.

In India's population program in the mid-1970s, the incentives to accept family-size limitations began to merge into coercion. At that point, public and political opposition grew until the program became the pivotal issue in the 1977 national election. The opposition Janata Party's defeat of Indira Gandhi and the Congress Party in that election led to a reversal of population policies.

There have been other, less dramatic public reactions to population policies, in response to overall policy or to such particular issues as the promotion of specific contraceptive methods. The use of Depo Provera in Bangladesh created public controversy because it had not been licensed in the United States and some other Western countries, and the administrators of the population program became vulnerable to charges of dispensing unsafe technology.

Population programs are most likely to generate bureaucratic conflict and resistance when leaders' commitment to limiting population growth is weak. The Kenya case cited earlier is an example, but the problem is widespread. International donors—public and private—as well as the United Nations, have put a high priority on population programs. A country that has no population program appears reluctant to modernize in the eyes of the world. Donors have ubiquitously offered aid for the adoption of such programs, often tying the flow of aid to their existence. As a consequence, policy elites may give formal support to programs that are not backed by their serious commitment to population policy objectives. Such soft commitment is readily apparent to program managers. Indeed, convincing a bureaucracy to accept the goals of population limitation is not easy when support from the top for such policies is weak, as was the case in Kenya. Summarizing the Kenyan situation, Donald Warwick (1982: 14) states, "Sensing ambivalence, indifference, and even hostility at the top, middle level administrators did not give their all to implementation." Without the necessary support of policy elites, bureaucratic commitment is unlikely and successful program implementation impossible. Ironically, however, in those few cases where top political leaders became highly committed to population control, there ensued public conflict when the programs were administered, as in India (noted earlier) and China.

Managers of population programs can succeed only if a broad segment of the public shares the program goals and adopts new forms of personal behavior. The managers are implementing a program that will be sustained only if they can persuade the public and then meet the demand they have created. A program's sustainability is tested, therefore, at the juncture between administrator and individual or between the country's stated priorities and individual incentives.

## Mobilizing resources to implement and sustain policy reform

If decisionmaking elites genuinely desire population limitation programs, then they must consider the resources that can be mobilized in support of their objectives. Sustainable programs depend more upon bureaucratic capacity and compliance than upon the stability of the regime in power. Unless the issue becomes a matter of public controversy, the focus will be on gathering political and bureaucratic resources to support the program. Resources, although almost always scarce, are not necessarily fixed, and can be mobilized through the conscious and concerted efforts of policy elites and managers.

The answers to two questions may suggest political resources that can be important in introducing and sustaining a new population program. First, how legitimate is the regime? If the government has strong, broad-based legitimacy, it is much more likely to be able to generate public support for any program. In undertaking a new population program, a government should assess its bases of support. Second, how autonomous is the government? If it depends on one or two extremely powerful interest groups, then the issue of how those groups will react is critical. Alternatively, if there is a consensus of elite groups in support of the government and its program, then mobilizing those groups is feasible. The response of the press, the financial community, the private sector, the military, and religious leaders can make a critical difference to the outcome of a population limitation policy. If the government can rely on these groups for support or obtain some assurance that they will not be actively opposed, then prospects for sustaining the program and influencing the public are greatly enhanced. An example of such an elite consensus occurred in Indonesia in the early 1970s, when program managers made great efforts to incorporate all major elite groups in support of population limitation policy.

Elite groups have, in some instances, become committed to population limitation in the absence of a broad social demand for such reform.

These groups, often comprised of doctors, educated women, and intellectuals, become advocates for population activities. They frequently align themselves with foreign private or public donors to become an influential coalition. Such a coalition may be powerful enough to come to the attention of political leaders and to play a role in the adoption of population programs, but not be sufficiently powerful to persuade their government to give new policies unequivocal backing. Partial or limited support from program managers during implementation will be the likely result, and programs will be born and financed, but not effectively implemented.

An analysis of the political resources needed to sustain a population measure cannot be limited to consideration of those who support or oppose the reform, but must also include an assessment of the degree to which support (and opposition) can be mobilized, how powerful each group is likely to be, and the way in which information about the program reaches people. It is also important for population programs to reach beyond national elites to rural areas and local elites (see Lindenberg and Crosby, 1981).

In addition to political resources, bureaucratic, financial, and technical resources must be mobilized by policy elites and public managers in order to sustain the programs. Although budgetary resources are always tight, underspending is a recurrent problem in some developing countries. A manager can usually work to have these funds shifted to an important program (see Ames, 1987). In addition, special accounts or funds may be available to knowledgeable and influential policy elites. Foreign aid is also frequently available to support population limitation activities. Many donors have often been able to obtain funds in addition to their country budgets when they have perceived themselves to be leveraging new or more effective programs.

When population limitation programs show potential for success, the capacity to generate more practical support for them is considerable. Control of budget, personnel appointment and promotion, and services ranging from transportation to purchasing are important elements of bureaucratic power that, when available, can make a population program attractive.

The technical capacity to deliver family planning information and services must be assessed by the officials considering the introduction of a program. Analysis of social attitudes and practices can lead to important insights about the solution of problems. As the program moves into practice, people with technical and paramedical skills are also needed.

In the cases we reviewed, we found considerable evidence of effective political management of the introduction of population policy—that is, pushing a policy or program into law. However, we found equal evidence of the failure to implement population policies successfully, apparently the result of a failure to generate extensive political support for the program or to plan strategically for its accomplishment.

Decisionmakers must make a long-term commitment to support the policy they promote and to develop an effective strategy that weds national priorities to individual incentives, both in the bureaucracy and among the groups expected to adopt new forms of population control. In these two areas, the gaps between rhetoric and practice are largest.

## Conclusion

Given the importance of population limitation policies and the difficulty of implementing them, are there grounds for hope in achieving more effective programs? We believe there are. Policy elites are of critical importance to the setting of agendas and decisionmaking about reform. These elites should become equally involved in designing implementation strategies and mobilizing the resources to carry them out. Unfortunately, policymakers often assume that once the decision has been taken, it will be carried out automatically by program managers and be readily accepted by at least part of the population.

The cases reviewed here indicate that these assumptions are unrealistic. Broad-based commitment to the goal of population limitation cannot be assumed just because population policy is adopted. Organizations, program managers, and administrators will not necessarily comply with the directives of new policies, nor will they necessarily have the skills and capacity to do so. Similarly, people will not necessarily feel an incentive to adopt birth control methods just because they become available.

A strategy for implementation can respond to the unwarranted assumption in several ways: (1) through political analysis of sources of support and opposition; (2) through the introduction of incentive systems, participation in decisionmaking, and developing managerial skills; and (3) through education, the development of responsive behavior on the part of bureaucrats and organizations, and through greater community involvement in service-delivery programs. National and individual priorities must be more closely related if population limitation programs are to succeed. This will only happen when national policymakers begin to understand and incorporate the values, points of view, motivations, and concerns of both bureaucratic agents and individual citizens.

## Notes

1 For our purposes, a reform initiative is any instance in which such a change is advocated; we pass no judgment on the efficacy or intent of any particular change being advocated.

2 We use the terms agenda and agenda-setting to refer to situations in which there is clear official acknowledgment that a problem exists and that its solution is to be sought through government action.

3 The term policy elites is used throughout this chapter to refer to political and bureaucratic officials who have decisionmaking responsibilities in government and whose decisions become authoritative for society. The term is used interchangeably with "decisionmakers" and "policymakers."

4 The framework is presented in Grindle and Thomas (1991). It was originally developed in Grindle and Thomas (1989). A discussion of implementation as it relates to our framework is found in Thomas and Grindle (1990).

5 As we have defined it, a situation of crisis exists when: (1) decisionmakers believe a crisis exists; (2) there is a consensus among them that the situation of crisis is real and threatening; and (3) they believe that failure to act will lead to even more serious economic and/or political upset in the immediate future. Situations of crisis or politics-as-usual are therefore defined by the perceptions of those in government who are responsible for particular policy areas.

6 The notion of "embedded orientations" is used by Bennett and Sharpe (1985) to explain historical biases in state policies. They use it in reference to states; we expand the notion here to apply to individuals.

7 For more general aspects of cultural factors in population policy, see Caldwell and Caldwell (1987), Korten (1975), and Weeks (1988).

8 The notion of chosen or pressing problems is Albert Hirschman's. He distinguishes between issues that are pressed on government officials by those outside government and those that are selected "out of thin air" by policymakers. See Hirschman (1981).

9 The only exception to this general pattern among the cases we reviewed was Kenya, where population policy choice appears to have been linked to an economic crisis: A population policy announcement was given as a sign to Western investors and creditors that the government of Kenya was taking appropriate steps to spearhead the country's economic development. Even in Kenya, however, the sense of urgency grew from the economic situation and was focused on generating this sign of responsibility, not on limiting population growth. Hence, the importance attributed to introducing a policy did not translate into a sense of urgency about pursuing it.

## References

Ahmad, Zakaria Haji, and Siew-Nyat Chin. 1975. "Organizing for population change in Malaysia," in *Policy Sciences and Population*, eds. Warren F. Ilchman, Harold D. Lasswell, John D. Montgomery, and Myron Weiner. Lexington, MA: Lexington Books, pp. 281–298.

Alba, Francisco, and Joseph E. Potter. 1986. "Population and development in Mexico since 1940: An interpretation," *Population and Development Review* 12, 1: 47–75.

Ames, Barry. 1987. *Political Survival: Politicians and Public Policy in Latin America*. Berkeley: University of California Press.

Bennett, Douglas C., and Kenneth E. Sharpe. 1985. *Transnational Corporations versus the State: The Political Economy of the Mexican Auto Industry*. Princeton: Princeton University Press.

Caldwell, John C., and Pat Caldwell. 1987. "The cultural context of high fertility in sub-Saharan Africa," *Population and Development Review* 13, 3: 409–437.

David, Henry P., and Mechai Viravaidya. 1986. "Community development and fertility management in rural Thailand," *International Family Planning Perspectives* 12, 1: 8–11.

Finkle, Jason L. 1972. "The political environment of population control in India and Pakistan," in *Political Science in Population Studies*, eds. Richard L. Clinton, William S. Flash, and R. Kenneth Godwin. Lexington, MA: Lexington Books, pp. 101–128.

Frank, Odile, and Geoffrey McNicoll. 1987. "An interpretation of fertility and population policy in Kenya," *Population and Development Review* 13, 2: 209–243.

Greig, James D. 1973. "Mauritius: Religion and population pressure," in *The Politics of Family*

*Planning in the Third World*, ed. T. E. Smith. London: George Allen and Unwin, pp. 122–167.

Grindle, Merilee S. (ed.). 1980. *Politics and Policy Implementation in the Third World*. Princeton: Princeton University Press.

———, and John W. Thomas. 1989. "Policy makers, policy choices, and policy outcomes," *Policy Sciences* 22: 213–248.

———, and John W. Thomas. 1991. *Public Choices and Policy Change: The Political Economy of Reform*. Baltimore: The Johns Hopkins University Press.

Hardee-Cleaveland, Karen, and Judith Banister. 1988. "Fertility policy and implementation in China, 1986–88," *Population and Development Review* 14, 2: 245–286.

Hirschman, Albert O. 1981. "Policymaking and policy analysis in Latin America—a return journey," in Albert O. Hirschman, *Essays in Trespassing: Economics to Politics and Beyond*. Cambridge: Cambridge University Press, pp. 142–166.

Hopkins, Thomas J., and Roble J. Siamwiza. 1989. "Converging forces: Processes leading to the rethinking of population policy in Zambia," in *Issues in Zambian Development*, eds. Kwaku Osei-Hwedoe and Muna Ndulo. Roxbury, MA: Omonena Press.

Hull, Terence, and Valerie Hull. 1973. "Fiji: A study of ethnic plurality and family planning," in *The Politics of Family Planning in the Third World*, ed. T. E. Smith. London: George Allen and Unwin, pp. 168–216.

Knodel, John, Napaporn Havanon, and Anthony Pramualratana. 1984. "Fertility transition in Thailand: A qualitative analysis," *Population and Development Review* 10, 2: 297–328.

Korten, David. 1975. "The importance of context in population policy analysis," in *Policy Sciences and Population*, eds. Warren F. Ilchman, Harold D. Lasswell, John D. Montgomery, and Myron Weiner. Lexington, MA: Lexington Books, pp. 139–148.

Lee, Eddy, Michael Ong, and T. E. Smith. 1973. "Family planning in West Malaysia: The triumph of economics and health over politics," in *The Politics of Family Planning in the Third World*, ed. T. E. Smith. London: George Allen and Unwin, pp. 256–290.

Lindenberg, Marc, and Benjamin Crosby. 1981. *Managing Development: The Political Dimension*. West Hartford, CT: Kumarian Press.

Lopez, Maria Elena, and Ana Maria Nemenzo. 1976. "The formulation of Philippine population policy," *Philippine Studies* 24: 417–438.

Maru, Rushikesh, Nirmala Murthy, and J. K. Satia. 1986. "Management interventions in established bureaucracies: Experiences in population-program management," in *Beyond Bureaucracy: Strategic Management of Social Development*, eds. John C. Ickis, Edilberto de Jesus, and Rushikesh Maru. West Hartford, CT: Kumarian Press, pp. 155–181.

Ness, Gayl D., with Hirofumi Ando. 1971. "The politics of population planning in Malaysia and the Philippines," *Journal of Comparative Administration* (November): 296–329.

Pyle, David. 1985. "Indonesia: East Java family planning, nutrition, and income generation project," in *Gender Roles in Development Projects*, eds. Catherine Overholt and Mary Anderson. West Hartford, CT: Kumarian Press, pp. 135–162.

Radel, David. 1973. "Kenya's population and family planning policy: A challenge to development communication," in *The Politics of Family Planning in the Third World*, ed. T. E. Smith. London: George Allen and Unwin, pp. 67–121.

Snodgrass, Donald. 1979. "The family planning program as a model of administrative improvement in Indonesia," *Development Discussion Paper* No. 58. Cambridge, MA: Harvard Institute for International Development.

Thomas, John W., and Merilee S. Grindle. 1990. "After the decision: Implementing policy reforms," *World Development* 18, 8: 1163–1181.

Thomson, George G., and T. E. Smith. 1973. "Singapore: Family planning in an urban environment," in *The Politics of Family Planning in the Third World*, ed. T. E. Smith. London: George Allen and Unwin, pp. 217–255.

Warwick, Donald P. 1982. *Bitter Pills: Population Policies and Their Implementation in Eight Developing Countries*. New York: Cambridge University Press.

———. 1986. "The Indonesian family planning program: Government influence and client choice," *Population and Development Review* 12, 3: 453–490.

Weeden, Donald, Anthony Bennett, Donald Lauro, and Mechai Viravaidya. 1986. "An incentives program to increase contraceptive prevalence in rural Thailand," *International Family Planning Perspectives* 12, 1: 11–16.

Weeks, John R.1988. "The demography of Islamic nations," *Population Bulletin* 43, 4: 4–6.

Whyte, Martin King, and S. Z. Gu. 1987. "Popular response to China's fertility transition," *Population and Development Review* 13, 3: 471–493.

# THE POLITICAL ENVIRONMENT OF FAMILY PLANNING PROGRAMS

# The Politics of Fertility in Africa

Omari H. Kokole

By 1989, the population of the African continent was estimated at 660 million (UNESCO, 1989). This figure represents nearly 10 percent of the world population. Considering that Africa occupies about a quarter of the world's land area, that is a modest figure. However, Africa's is a rapidly expanding population. The current average annual growth rate of 3 percent is not only the highest on the globe (United Nations, 1990), but also, according to the World Bank (1989: 40), "the highest seen anywhere, at any time, in human history."[1] This pace of population expansion is much greater than the average for the entire third world (including Africa; Asia—excluding Japan, Hong Kong, South Korea, and the People's Republic of China; and Latin America), which is slightly more than 2 percent. It also far outstrips the annual growth rate of world population, estimated at 1.7 percent for 1990–95 and the 0.5 percent rate of expansion for the developed societies of the Northern hemisphere (UNFPA, 1992). Should current trends continue, Africa's population would double every two decades. Meanwhile with the exception of a handful of countries (for example, Botswana, Mauritius, Zimbabwe, and Kenya) Africa's economies have been seriously ailing. In 1987, Africa had a total gross domestic product (GDP) of US$135 billion, roughly the equivalent of the GDP of Belgium, whose population is barely 10 million (World Bank, 1989). The collective external debt of the continent exceeds US$300 billion.

Averaging 3.4 percent a year since 1961, Africa's aggregate economic growth has been only a fraction above population growth. The effects of this poor economic performance are exacerbated by other problems, including desertification, deforestation, rural–urban migration, infrastructural decay, unsanitary living conditions, and lack of clean water. The quality of life has steadily deteriorated in the postcolonial era.

Reducing fertility rates would help many countries on the continent to advance on the path to development; yet, family planning options are limited in much of Africa. In this article, the political and cultural constraints that have impeded family planning programs in Africa are identified and analyzed. Particular attention is paid to the interplay between indigenous values and Western values, within a range of cultural, political, military, and economic variables.

## Cultural values and procreation

The cultural factors that influence population trends in Africa include indigenous values and deep-rooted beliefs that place a high premium on begetting children. Religious beliefs and marriage customs of the various communities of Africa are particularly influential. As Caldwell and Caldwell (1987) have observed:

> Sub-Saharan Africa may well offer greater resistance to fertility decline than any other world region. The reasons are cultural and have much to do with a religious belief system that operates directly to sustain high fertility but that also has molded a society in such a way as to bring rewards for high fertility. (p. 409)

The belief that there is life after death and that the length of that life depends in part on the number of children one begets is fairly widespread. The passport to the immortality of any adult individual, the ultimate guarantee to being remembered as an ancestor in the hereafter, is by, and through, one's progeny. Again, according to the Caldwells:

> The essence of the traditional belief system [i.e., indigenous culture] is the importance attributed to the succession of the generations, with the old tending to acquire even greater and more awe-inspiring powers after death than in this world and with the most frequent use of those powers being to ensure the survival of the family of descent. (Caldwell and Caldwell, 1987: 409)

These beliefs about parental immortality are in turn linked to issues of infant mortality. Africa is a continent of "abundant life but speedy death" (Mazrui, 1986: 41). The high infant mortality rates create an incentive for having as many children as possible because it is uncertain how many, if any, will survive. The infant mortality rate for Africa, though declining, remains the highest in the world. In 1992, there were 94 deaths of African infants aged one year or younger per 1,000 live births, compared with 63 deaths per 1,000 live births for the world as a whole; 12 for the industrialized countries; and 70 for the third world as a whole (UNFPA, 1992).

The belief in the importance of having many children is fortified in Africa by economic, political, and psychological realities. African children have traditionally contributed to their families' economic well-being by herding cattle and performing a variety of farm and domestic chores. This remains basically true even when the children go to school. Many children who attend urban schools live in rural areas and are expected to help with domestic and related work after school hours. Their contribution increases during their three-month vacations from school.

For most Africans, the rural–urban gulf is not wide. A rural–urban continuum exists whereby ties to the land remain strong. Clearly, having more children means having more potential workers. If funds for educating children are scarce, the brighter children may continue their formal education while the less able remain behind to work. Others may seek employment opportunities further afield. Often, some members of extended families live in the cities and others in the countryside.

Even in urban settings, large families may be advantageous. Caldwell discovered that urban professional Yoruba families in Nigeria profited from having numerous children through "sibling assistance chains." Once a brother or sister completes education and is gainfully employed, he or she often helps pay for younger brothers' and sisters' education and also helps them to find employment (Caldwell, 1982; Hartmann, 1987). In much of Africa, many city dwellers remit money periodically to their poorer relatives in the countryside, as do many Africans living and working in the developed world, especially in the West.

Having many children is also considered a form of insurance for parents in their old age. Since high infant mortality rates cause doubts about child survival, about how many children might live long enough to support their aged parents, having many children appears sensible.

The perceived need to have many children is one incentive for the practice of polygyny (Brown, 1981). In addition to bearing more children for the lineage, the co-wives augment the domestic labor force. Many a first wife in Africa encourages her husband to take other wives precisely for this reason:

> A Tanzanian woman with 13 children, recently interviewed by a journalist, explained why she was disappointed that her husband had refused, despite her urging, to take another wife. She felt her life had been unduly hard, because she had to do all the farming and food preparation herself, without the help of other wives or her husband, because that was not his role. As a result, she was undeniably bitter about her seemingly endless years of child-bearing and toil in the fields. (Valentine and Revson, 1979: 459)

Many African women accept frequent childbearing partly because otherwise their hold on their men would be jeopardized. Rejection or divorce on the grounds of infertility is commonplace. The greater the number of children, the greater the social prestige and respect for the mother. A wife's low fertility might suggest a man's impotence, and high fertility is therefore valued. Infertility and infecundity can also be interpreted as signs of ancestral, or even divine, displeasure.

## The politics of high fertility

Politically, there are advantages to having many children. In the centralized political communities of precolonial Africa, especially in the monarchies and empires, the leader aspired to many children partly in order to beget a suitable heir. Sons were favored in this regard, but daughters, too, were useful, especially in building alliances through marriage (Davidson, 1984; Caldwell, 1982; Molnos 1973). The larger the size of the family, the greater the prestige and status for the parents.

Because the royal houses became models for nobility and commoners throughout African society, this reproductive pattern was often widely imitated. But throughout Africa, in all sorts of communities, having a sizable number of children of both sexes has been desirable in part because the marriage of daughters reaps bridewealth (so-called bride price), which is useful when gaining a wife for one's son. The sons bring other advantages and benefits of their own.

If Africa's indigenous cultures and religions have been responsible for the birth of millions of children, the culture most relevant to the survival of African children has been both Western and secular—Western science and technology.

A closely related Western contribution has been the reduction of sterility as a result of modern medical facilities. As Romaniuk (1980) discovered in Zaire, modernization may increase natural fertility when it reduces large-scale pathological infertility (for example, from venereal diseases) among societies that do not practice birth control. Not only have antibiotics curtailed infertility, but Western cultural influences have helped to undermine the indigenous practice of child spacing (that is, of extended postpartum abstinence and protracted breastfeeding) (Bongaarts, 1981; Larsen, 1989).

## Ethnic competition and population politics

Politically, modern Africa is in one fundamental sense a product of European imperialism and colonialism. Virtually all African countries are artificial societies, the boundaries of which were arbitrarily drawn without reference to the continent's long precolonial history and cultures. Often these made-up frontiers split people who had a common heritage even as they united tribes with little in common.

In the postcolonial era, the colonial boundaries have created and intensified ethnic conflicts and contests. Confrontations often occurred that were related to competition for political power or scarce economic resources. Frequently, the demographic balance between different ethnic or cultural groups enclosed within the same boundaries has become a political issue. Population size often determines the sharing of resources and even power. In Nigeria, the most populous of all African countries, the comparative size of ethnic populations has been a major political issue in both the colonial and postcolonial eras. Adepoju notes:

> [T]he census would form the basis for revenue allocation, provision of amenities and, more important, representation in the forthcoming elections into regional and federal legislatures in 1964. . . . Population issues . . . precipitated the constitutional crisis in the country in 1962; [they] played a major role in the crisis in the old Western Nigeria in 1965; [were] largely responsible for the military takeover in 1966; contributed greatly to the fall of Gowon's regime in 1975 and still [loom] large in the minds of Nigerians with the recent demand for the creation of even more states in the country; and revenue allocation among existing states soon after the return to civilian rule. (Adepoju, 1981: 29–30)

Even today in Nigeria, census figures remain highly political. The British imperial policy of indirect rule tended to sharpen ethnic or "tribal" loyalty and thus increased the dangers of violent confrontation among the ethnic communities after independence. By institutionalizing ethnic pluralism in colonial Nigeria, the British set the stage for anarchy in the future. While British colonial rule lasted, the struggle for political independence itself generated some degree of unity among the various ethnic groups. But after independence, the fragile unity dissipated.

Before the civil war (1967–70), Nigerians passionately debated the population tallies of various ethnic groups and their implications for the electoral process. The 1958 constitutional conference had provided for the Federal Assembly to be elected on the basis of population figures (Mazrui and Tidy, 1984). This stipulation caused each community to regard as suicidal absurdity any effort to reduce its fertility rate. That the three largest ethnic groups in the country (the Hausa-Fulani in the north, the Yoruba in the West, and the Igbo in the East) had been the dominant actors in the national life of the country seemed to validate the potency of numerical strength. Not surprisingly, as Elaigwu (1984: 9) has observed, "The South . . . feared Northern political domination by population—the *tyranny of* population by the North."

Odumegwu Emeka Ojukwu proclaimed the predominantly Igbo Republic of Biafra on 30 May 1967. The futile attempt at secession triggered a civil war that dragged on for three years and cost thousands of lives on both sides (De St. Jorre, 1972; Akpan, 1971). Civilians bore the brunt of hostilities and died in far larger numbers than the military.

During the armed conflict in Nigeria, the idea of family planning was distant and academic; instability and domestic anarchy are not conducive to purposeful fertility planning. The formation of the National Population Council as the highest political body on population activities in Nigeria had to await the end of the civil war in 1970 (Adepoju, 1981).

Another former British colony that suffered devastating consequences of indirect rule is Uganda. Indirect rule led to the preservation of some strong indigenous institutions, especially the monarchies of the southern part of the country. These kingdoms included Buganda, Bunyoro, Ankole, and Toro.

Ethnically, Uganda is a plural society. Although the country has many ethnic groups, the great political divide in recent times has been that between the "Nilotic North" and "Bantu South." Demographically the Bantu groups of the south outnumber the Nilotes of the north by a ratio of four to one. Bantu-speakers represent more than 70 percent of Uganda's population. Of these, the Baganda are the largest ethnic group, comprising about 20 percent of the total national population (Kokole, 1991). The greatest rivalry in postcolonial politics has been between Buganda, the largest and wealthiest region since the colonial era, and the precolonially decentralized North (Kokole and Mazrui, 1988).

The method of European penetration into Uganda and the way the Europeans transformed the economy favored the Bantu. To this day, a disproportionate number of educated Ugandans are Bantu. The Bantu have dominated the modern sector of society, occupying most of the high academic, judicial, bureaucratic, religious, and other prestigious positions. British colonial activities created a situation in which modernity, economic power, and prosperity were concentrated in southern Uganda. The traditionally stateless north was left far behind.

On the other hand, the army and security forces that the British created were recruited overwhelmingly from the ethnic groups of the north. Just as the British had preferred the "martial" Gurkhas of Nepal and Northern India to other Asians to serve as their soldiers, in Uganda the British tended to recruit the "martial" tribes of northern Uganda. Although their basis for distinguishing between "martial" and "nonmartial" groups remains questionable, the British behaved as if the Nilotic northerners in Uganda were, indeed, martial material, and as if the Bantu southerners were not (Mazrui, 1975). As a result, until Yoweri Museveni's historic takeover in January 1986, the armed forces, the police, the paramilitary corps, and the prison system were all dominated by northerners.

The political developments of postcolonial Uganda have, to a large extent, been shaped by this separation of economic and military power. To be sure, the lines of cleavage have not always been drawn between the Bantu on one side and the Nilotes on the other. But most informed observers concur that there is a basic north/south divide in the political economy of independent Uganda (Mutibwa, 1992; Omara-Otunnu, 1987; Gingyera-Pinycwa, 1978; Kokole, 1991; Mazrui, 1975). This situation has sharpened ethnic consciousness, deepened sensitivity about ethnic populations, and dampened interest in reducing fertility rates. The chronic instability and military conflict in Uganda have undermined the formulation of a population policy.

The impact of British divide-and-rule tactics on fertility decline and population planning in Africa should not be overemphasized. Ethnic competition and tension were among the saddest legacies of the colonial era, even where different policies were pursued by the British, as in Kenya, or by the Belgians in what is now Zaire, Burundi, and Rwanda, or by the French in their colonies.

In Kenya, the equivalent of the Baganda of Uganda is the Kikuyu, the single largest ethnic group in Kenya. Kikuyu domination of the polity and the economy has been somewhat loosened under the presidency of Daniel Arap T. Moi, but it remains substantial. By living near the capital, Nairobi, the Kikuyu enjoy relatively easy access to educational and economic opportunities.

The second largest ethnic group in Kenya is the Luo, who play a major, if lesser, role in the politics of the country. In the postcolonial era, the Kikuyu and Luo have been political rivals, acutely aware of their comparative ethnic sizes (Wilks, 1970). Also, as Goliber (1985: 31) points out, "the smaller tribes [in Kenya] are sensitive to their relative strength in the political order, and this is related to numbers." Clearly, the struggles for ethnic hegemony or ethnic equality in Kenya, as in most African countries, discourage any effort to reduce fertility rates. Unless family planning programs are perceived to affect all ethnic groups equitably, their efficacy and success will remain uncertain (Mott and Mott, 1980).

Kenya established its national family planning program in 1968, the first country in black Africa to do so (Caldwell and Caldwell, 1987). More than two decades later, the program's success remains elusive. Currently Kenya's annual population growth rate of 4.1 percent (approximately eight births per woman) is the highest in the world (Goliber, 1985). Kenya's population, estimated at 24 million in 1991, is expected to increase to 35 million by 2000 and to 79.1 million by 2025 (UNFPA, 1992).

Why have more than two decades of governmental family planning efforts in Kenya failed? A fundamental factor is the resilience of indigenous African culture discussed earlier and the value that culture places on having many children. Second, the country's charismatic founder—Jomo Kenyatta—who ruled the country as its first president for 15 years (1963–78), was a cultural nationalist (Kenyatta, 1971; Wilks, 1970). Although he did not oppose population policy, he did not endorse it strongly either. Third, the Western preference for small families, to which Africans are exposed through the modern educational system, the media, and travel, has yet to impress many Kenyans. The major role that the United States plays in the population policy exercise may make the ideas therein appear artificial and foreign. Family planning activity in Kenya did not excite and engage people partly because its principal advocates were foreigners.

In addition, substantial foreign aid provided for fertility-change programs may have been diverted to other purposes by local officials who did not support or believe in population policy. People's loyalty remained focused not on the new and artificial entity "Kenya," but on their respective ethnic communities.

In francophone Africa, the colonial experience helped to strengthen a preexisting pronatalist orientation. In this case, indigenous African culture and Western (French) influence—including the pervasive influence of the Roman Catholic Church—converged. French pronatalist influence and legislation has in fact survived the colonial era (Caldwell, 1973). Partly as a result of these factors, francophone Africa lags behind the rest of the continent in adopting policies intended to curtail fertility. For example, "By 1986 only Chad, Côte d'Ivoire, Gabon, Guinea-Bissau, and Mauritania were pronatalist or gave little support to family planning" (World Bank, 1989: 71). Apart from lusophone Guinea-Bissau, all the countries mentioned above are francophone. The bulk of francophone Africa lies in the western part of the continent, whence most of the victims of the trans-Atlantic slave trade came. Demographically the region suffered. A familiar profertility thesis, coming especially from the level of political elites, is that an imperative exists to compensate for the population decimation wrought by the slave trade.

Another profertility argument in Africa is shared by many African-Americans, namely that the effort to reduce fertility, especially when encouraged or led by Northerners and institutions they control, such as the World Bank and the International Monetary Fund, is part of a conspiracy to undermine the political weight of people of color (Green, 1982; Armar, 1983; Teitelbaum, 1992–93). Clearly, those who lack numbers are less influential than those whose numbers are great. Arguably, the cause of black South Africans would have been less weighty and compelling had they been in the minority. Having numbers on its side is one of the most powerful weapons of any dispossessed majority.

## Dual societies and competitive fertility

Although ethnic competition and conflict are widespread in contemporary Africa, ethnically dual societies, which consist of two nonintegrated communities that are often split politically, like Rwanda, Burundi, Sudan, and Zimbabwe, pose special problems (see Mazrui, 1981).

In the case of Rwanda, the ethnic divide has been between the Hutu and the Tutsi. During most of the twentieth century, the minority Tutsi (15 percent) governed virtually as feudal overlords. Tutsi chiefs managed the natural and human resources of the country. Majority Hutu political response began in 1967 when the numerically small Hutu Catholic elite appealed for the termination of Tutsi domination and for fundamental reforms. This elite issued the *Hutu Manifesto*, which reaffirmed the commitment to redress the balance in Hutu–Tutsi relations (Mazrui and Tidy, 1984). The Tutsi responded by further consolidating their privileges and authority. Following the death of the Tutsi king, Mwami Matara III, in July 1959, a merciless Tutsi clan decided to put the presumptuous Hutu in their place once and for all. A massacre and a wave of intimidation of the Hutu followed. Hutu–Tutsi conflicts and fighting continued after independence. In the preindependence elections the Hutu majority party, *Parmehutu*, won a landslide victory, and the Hutu leader Grégoire Kayibanda (a Catholic by faith and journalist by training) became the first president of independent Rwanda on 1 July 1962.

Accustomed to dominating the Hutu for so long, most Tutsi refused to concede the Hutu electoral victory. Instead they formed a guerrilla organization, enigmatically named Cockroach, which they based in neighboring and Tutsi-dominated Burundi. When Cockroach attempted in December 1963 to invade Rwanda and topple the Hutu regime, the consequences for the Tutsi were dire. Retribution for the failed invasion was swift and brutal; it led to hundreds of thousands of Tutsi deaths and to Tutsi emigration. Further killings of Tutsi in Rwanda occurred in the early 1970s in response to killings of Hutu by Tutsi soldiers in neighboring Burundi. The ensuing anarchy and violence led the army commander, General Juvenal Habyarimana, to take power in a 1973 coup and to outlaw Parmehutu.

In its rhetoric at least, the Habyarimana regime tried to downplay ethnicity in Rwanda. However, Hutu–Tutsi tensions and suspicions in Rwanda remain and have helped to hinder the development of a population policy or organized family planning programs (Abate, 1978). Dual societies, like Rwanda's, tend to sharpen and aggravate ethnic conflicts, and one casualty is population policy.

Burundi is also a dual society. There, the Tutsi dominated the Hutu politically, economically, and socially from the precolonial era, through the colonial interlude, and into the postcolonial years. As in Rwanda, the colonial experience exacerbated ethnic conflicts between the two groups. Tutsi chiefs in Burundi commanded local administration and the land. The minority Tutsi (15 percent) had far greater access to (modern) mission education than did the Hutu majority (80 percent).

But the Tutsi aristocracy had also led the struggle for independence in Burundi. Although the Tutsi had at first involved the Hutu in the nationalist movement—the Union for National Progress (UPRONA)—Tutsi tribal fanatics later succeeded in sweeping out the Hutu from UPRONA and in consolidating Tutsi hegemony.

When Hutu candidates won a majority in the 1964 elections, Mwami (King) Mwambutsa II (1962–65) vehemently refused to appoint a Hutu as prime minister. What followed was an attempted military coup by the Hutu. Although the coup failed, Mwami Mwambutsa fled to Europe. The vacuum he left behind was filled by a group of Tutsi loyalists who captured power and eliminated the Hutu from both the civil administration and the military.

The Tutsi were neither a homogeneous nor a cohesive clique. Differences among them brought about a military coup in mid-1966 that ushered into power yet another Tutsi, Colonel Michel Micombero. Later that year, the colonel deposed the last king of Burundi, Mwami Ntare V. Under Micombero, Tutsi dominance remained intact. UPRONA became a weapon of ethnic victimization. The movement engaged in recurrent expulsion of Hutu from the party itself, the bureaucracy, and the security and armed forces. In response the Hutu rebelled in early 1972, partly in protest against the murder of leading Hutu figures. Approximately 1,000 Tutsi were killed in the confrontation. In retaliation the Tutsi-packed army carried out a "selective genocide," murdering all the educated Hutu it could find. The total number of Hutu who lost their lives in this conflict is estimated at 200,000. Another 70,000 Hutu fled as refugees to neighboring countries, mainly Tanzania.

Although since Micombero's overthrow in 1976 Burundi has had other leaders, the country remains a Tutsi oligarchy. Before 1992 it was a single-party republic. The government of Burundi has yet to develop a national population policy (United Nations, 1987). Clearly, this failure is related to the Hutu–Tutsi conflict. Perhaps a related variable is the role of Roman Catholicism in the francophone countries of Burundi and Rwanda. This Western faith is strong in both dual societies.[2] Most postcolonial leaders in the two countries, though not necessarily pious or religious, were Roman Catholic. Just as the Church played a role in sustaining high birth rates in Latin America, it has condoned high fertility rates in Burundi and Rwanda, and discouraged national population policy.

Ethnic dualism in Zimbabwe concerns the divide between the Shona majority (70 percent of the population) and the Ndebele minority (about 15 percent). In their simultaneous and protracted armed struggle for black majority rule, the two communities, despite having a shared enemy, fought separately most of the time. The Shona formed the bulk of the Zimbabwe African National Liberation Army (ZANLA) under Robert Mugabe's leadership. The Ndebele on the other hand were the backbone of the Zimbabwe People's Revolutionary Army (ZIPRA). As Mazrui has observed:

> ZANLA and ZIPRA fighters were under separate commands, and seemed to have distinctive ethnic loyalties. Political leadership and military loyalty were thus ethnicized. (Mazrui, 1981: 8)

Even after they had won the struggle for black majority rule, the Shona and Ndebele remained separate politically. Leading a dual society, Robert Mugabe's government has been cautious in its concerns about a population policy.

Having lost many of its sons and daughters in the struggle for black majority rule, the Zimbabweans were also emotionally opposed to measures that would limit the replenishment of their population. Furthermore, the enemy, the white minority racists, had been the first to introduce the idea of family planning into the country. Accordingly, family planning was perceived as part of the white racists' grand design to prolong minority rule and deny blacks their birthright. White racism in Zimbabwe and elsewhere on the continent gave population planning a bad name.

Nevertheless, postcolonial and black-ruled Zimbabwe is considered to have one of the strongest family planning programs in Africa. Admittedly, as the result of adverse economic and ecological circumstances, the program is currently under stress. But many more Zimbabweans now approve of contraception than was the case before independence in 1980. According to the World Bank (1989: 70), "Zimbabwe and Botswana are the leaders in family planning in sub-Saharan Africa. Their programs are available to most citizens. Knowledge of modern contraception is widespread, and levels of modern contraceptive use—in 1988, 36 percent in Zimbabwe and 32 percent in Botswana—are the best in Africa." Fertility decline in Southern Africa is, surprisingly, outstripping other parts of the continent. Why has this shift occurred? The relevant factors include vigorous family planning programs, relatively high incomes, and increasing education of the population.

The Sudan is a dual society in a special sense. The country is fragmented between a predominantly Muslim-cum-Arab North and a basically non-Muslim-cum-black South. These differences are reinforced by the regional factors implicit in the North–South divide. The Sudan's duality involves, therefore, both religious polarization and geographic separation.

The civil war that has raged in the Sudan for almost a quarter century (from 1955 to 1972 and from 1983 to the present) is one of the longest-lasting civil wars in postcolonial Africa. In the first phase of the conflict, the black Africans in the South wanted to secede and create an independent state. The grievances of the Southerners were by no means primarily religious, although religious differences were involved. The Southerners were resisting their overall domination by the North for which the British had set the stage during the colonial period (Kokole, 1984). The British had governed the South separately from, and unequally with respect to, the North. After independence, the Northerners treated the Southerners poorly and with immense cultural condescension (Wai, 1981).

The Family Planning Association of the Sudan was established more than 25 years ago; it became an International Planned Parenthood Federation affiliate in 1971 (Nortman, 1982). But because of the civil war in the South, the Association's operations were restricted there, and its activities were limited to the North.

None of the several postcolonial governments of the Sudan has regarded the population growth rate of 2.8 percent as unsatisfactory. This complacent attitude is partly a function of the divide between North and South and of the armed conflicts it has generated. The Southerners, who comprise approximately 40 percent of the national population, were inclined to be suspicious and skeptical about a population policy "imposed by the North." The Southerners were not likely to look at population issues dispassionately. Although the Southerners are now less interested in secession and are more concerned about a new Sudan in which Southerners and Northerners would be treated equally, there is little doubt that comparative ethnic size will continue to be a politically sensitive issue and, therefore, a serious impediment to a national population policy.

## The demography of apartheid

Reference has already been made to Zimbabwe, but this country should be related to its regional context of Southern Africa. This portion of the continent was the last to gain independence from various European colonial rulers, and, where relevant, the last to have black majority rule introduced. Namibia, Africa's last colony, gained its independence in March 1990. However, the struggle for black majority rule in the Republic of South Africa continues. Since 1990, tremendous progress has been made in redressing the racial equation, but apartheid as a system of institutionalized racism has not been abolished altogether. The black person still does not have the right to vote; political power remains in white hands, even in 1993. The elections scheduled for April 1994 will mark the first time blacks are permitted to vote.

The colonization of Southern Africa by European powers and the peopling of the area by settlers of European origin were invariably accomplished by violence. The white man was able to assert his will in the region because of superior organization and military technology. The cost in terms of lives, property, and freedom for black Africans was considerable.

Southern Africa attracted a much larger number of white settlers than did other parts of the continent, partly because of its congenial climate, fertile soil, abundance of minerals, strategic location, and natural beauty. Wherever whites settled in large numbers in Africa they have never given up power without a fight. Consequently, the ensuing black–white struggles have been protracted and deadly, especially for the black majorities.

Until F.W. de Klerk's ascent to the presidency in 1989, the most obstinate of these white-controlled countries in the subcontinent was the racist

Republic of South Africa (RSA). In 1988, the republic had an estimated population of 35.1 million, increasing by 2.3 percent per year. Approximately 73 percent of this population was black, 16 percent was white, 9 percent was racially mixed "coloureds," and 3 percent was Asian. Blacks, with twice the annual birth rate of whites, outnumbered the latter by a ratio of 4.5 to 1 (Goliber, 1985).

According to the estimates of the US Bureau of the Census, there will be 34.5 million black South Africans by the year 2000. By then, the white population will stand at 5.8 million, a ratio of six blacks to one white. The projection for the year 2010 increases the disparity to 42.3 million black South Africans and 6.2 million whites, or a ratio of seven to one; the white sector would then be only 13 percent of the national population. All of these comparative statistics amount to a demographic nightmare for the apartheid regime (Johnson and Campbell, 1982; Goliber, 1985).

The creation of the Bantustans or "black homelands" (so-called independent black states) within the RSA was intended to "solve" this nightmare. The black majority was supposed to be squeezed into these small, remote, desolate, infertile, and least-developed parts of the country. The strategy of dislodging the bulk of the black masses from "white areas" into relatively distant black "homelands"—Ciskei, Transkei, Venda, Bophuthatswana, KwaZulu, Qwaqwa, KaNgwane, Ndebele, Lebowa, and Gazankulu (Study Commission, 1986)—in effect turned the black natives into strangers in their own land.

The same strategy would confirm the whites as sole residents of the wealthiest portions of the country. In such a situation of racial victimization and massive injustice, any attempt by white authorities in Pretoria to encourage blacks to have small families would arouse suspicions about the real motives of that policy. Indeed as Caldwell (1971: 70) observed, "[T]he mass white support [for family planning in South Africa] now developing undoubtedly bases its views on the fear of the political implications of a disproportionate increase in black numbers." Yet, in the past 20 years, the situation has changed considerably, and good private and governmental family planning programs now exist.

Because the strongest weapon that the blacks possess in the RSA is their relative numbers, and since these numbers form the basis for their claim to black majority rule, blacks would be understandably suspicious of any programs designed to reduce their population. The desire to keep their numbers high is reinforced by there being far more black than white casualties in the continuing violence generated by apartheid.

Before the 1990s, many of South Africa's black-ruled neighbors had strong political motives to resist policies that might result in reduced fertility rates. Most of these countries remain victims of South Africa's ruthless campaign of destabilization (until de Klerk became president and initiated reforms). When the South African government mounts raids and incursions into neighboring countries in "pursuit of terrorists," many innocent civilians are killed either directly by firepower or indirectly by the aftermath of these invasions. The activities of rebels supported by South Africa, such as the Mozambique National Resistance, the National Front for the Total Independence of Angola, and similar groups in a number of other countries, also result in much loss of life.

In addition to these destabilizing activities is the wider history of black–white armed conflict in the region and the bloody history of white settlement. Portugal, resistant to the end, held onto its African colonies ferociously until the mid-1970s before conceding independence to them. These events and developments cost thousands of black African lives in Angola, Mozambique, Zimbabwe, Botswana, and Zambia (United Nations, 1989; Hanlon, 1986). For all these reasons, the Southern African region is arguably the least inclined in all of Africa to adopt programs intended to reduce fertility rates. Given the problems and circumstances in these countries in the last 30 years, the quest for a national population policy could hardly be a priority.

## Conclusion

Demographic trends in Africa are alarming. The population of the continent is growing at an exponential rate. Meanwhile, economic and material conditions have not only failed to improve, but in many cases have deteriorated. Should these trends continue, the implications are grave indeed.

Africa's fertility rates are now the highest in the world. Many factors have contributed to this reality—political, religious, cultural, economic, and ecological. However, the contemporary politics of postcolonial Africa has been a dimension neglected by those who specialize in population issues. Demonstrably, political factors, especially ethnic rivalries in Africa's inchoate nation-states, impede the adoption of national policies intended to curtail human fertility.

Short of redrawing Africa's national boundaries to form new, more ethnically homogenous states (an unlikely prospect), possible solutions range from the decentralization of political power (distributing power to the various ethnic groups) to nation building used as a strategy to forge a national identity that would transcend ethnic loyalties. The expansion of secular modern education is a further possible solution.

## Notes

This article has benefited immensely from extensive discussions with Ali A. Mazrui on issues of population growth and policies in Africa.

1 Some of the oil-rich Gulf States in the Middle East have population growth rates greater than 5 percent, but these rates are basically the result of the influx of migrant workers.

2 According to the *Encyclopaedia Britannica*'s 1980 estimates for Burundi, Christians make up 79.5 percent of the population (the vast majority of whom are Roman Catholic); people of indigenous religions are 13.5 percent, nominal Christians, 6 percent; Muslims, .9 percent; and others .1 percent. For Rwanda, according to 1985 estimates, the figures were: Roman Catholics, 56 percent; Protestants, 12 percent; Muslims, 9 percent; and people of indigenous religions, 23 percent. See *Britannica Book of the Year, 1987*.

## References


Abate, Yohannis. 1978. "African population growth and politics," *Issue: A Journal of Africanist Opinion* 8, 4: 14–19.

Adepoju, Aderanti. 1981. "Military rule and population issues in Nigeria," *African Affairs* 80, 318: 29–47.

Akpan, N. U. 1971. *The Struggle for Secession*. London: Frank Cass.

Armar, A. A. 1983. "The Ghana National Family Planning Programme," in *Management Contributions to Population Programmes: Views from Three Continents*, ed. E. Satter. Kuala Lumpur: International Committee on the Management of Population Programmes: pp. 102–109.

Bongaarts, John. 1981. "The impact on fertility of traditional and changing child-spacing practices," in *Child-Spacing in Tropical Africa*, eds. Hilary J. Page and Ron Lesthaeghe. London: Academic Press, pp. 111–129.

*Britannica Book of The Year, 1987*. 1987. Chicago: Encyclopaedia Britannica, Inc.

Brown, Judith E. 1981. "Polygyny and family planning in sub-Saharan Africa," *Studies in Family Planning* 12, 8/9: 322–326.

Caldwell, John C. 1971. "Family planning programs and official policy decisions in Southern Africa." Unpublished manuscript.

———. 1973. "Family planning in continental Sub-Saharan Africa," in *The Politics of Family Planning in the Third World*, ed. T. E. Smith. London: George Allen and Unwin, pp. 50–66.

———. 1982. *Theory of Fertility Decline*. London: Academic Press.

———, and Pat Caldwell. 1987. "The cultural context of high fertility in sub-Saharan Africa," *Population and Development Review* 13, 3: 409–437.

Davidson, Basil. 1984. "The king and the city," from *Africa*, an eight-part television series, program no. 4. London: Mitchell Television/RM/Arts/ Channel Four Television Production.

De St. Jorre, John. 1972. *The Nigerian Civil War*. London: Hodder and Stoughton.

Elaigwu, J. Isawa. 1984. "Nigeria's federal system: Conflict and compromises in the political system," *University of Jos Postgraduate Open Lecture Series*, 1, 4, January.

Gingyera-Pinycwa, A. G. G. 1978. *Apolo Milton Obote and His Times*. New York: Nok.

Goliber, Thomas J. 1985. "Sub-Saharan Africa: Population pressures on development," *Population Bulletin* 40, 1: 3–47.

Green, E. 1982. "US population policies, development and the rural poor in Africa," *Journal of Modern African Studies* 20, 1: 45–67.

Hanlon, Joseph. 1986. *Beggar Your Neighbors: Apartheid Power in Southern Africa*. Bloomington: Indiana University Press.

Hartmann, Betsy. 1987. *Reproductive Rights and Wrongs: The Global Politics of Population Control and Reproductive Choice*. New York and London: Harper and Row.

Johnson, Peter D., and Paul R. Campbell. 1982. *Detailed Statistics and the Population of South Africa by Race and Urban/Rural Residence, 1950–2010*. Washington, DC: US Bureau of the Census.

Kenyatta, Jomo. 1971. *Facing Mount Kenya: The Tribal Life of the Kikuyu*. London and Nairobi: Heinemann Educational.

Kokole, Omari H. 1984. "The Islamic factor in African-Arab relations," *Third World Quarterly* 6, 3: 687–702.

———. 1991. "Uganda," *Encyclopaedia Britannica,* 15th edition, vol. 17. Chicago: Encyclopaedia Britannica, pp. 810–814.

———, and Ali A. Mazrui. 1988. "Uganda: The dual polity and the plural society," in *Democracy in Developing Countries,* vol. 2, eds. Larry Diamond et al. Boulder, CO and London: Lynne Rienner Publishers and Adamantine Press, pp. 259–298.

Larsen, Ulla. 1989. "A comparative study of the levels and the differential of sterility in Cameroon, Kenya and Sudan," in *Reproduction and Social Organization in Sub-Saharan Africa,* ed. Ron J. Lesthaeghe. Berkeley: University of California Press, pp. 167–211.

Mazrui, Ali A. 1975. *Soldiers and Kinsmen in Uganda*. Beverly Hills, CA: Sage Publications.

———. 1981. "Dual Zimbabwe: Toward avoiding political schizophrenia," *Issue: A Quarterly of Opinion* 11, 3/4: 5–12.

———. 1986. *The Africans: A Triple Heritage*. London: BBC Publications and Little, Brown.

———, and Michael Tidy. 1984. *Nationalism and New States in Africa: From about 1935 to the Present*. London and Nairobi: Heinemann.

Molnos, A. 1973. *Innovations and Communication,* vol.3. Nairobi: East African Publishing House.

Mott, Frank, and Susan H. Mott. 1980. "Kenya's record population growth: A dilemma of development," *Population Bulletin* 35, 3: 3–43.

Mutibwa, Phares. 1992. *Uganda Since Independence: A Story of Unfulfilled Hopes*. Trenton, NJ: Africa World Press.

Nortman, Dorothy L. 1982. *Population and Family Planning Programs: A Compendium of Data through 1981,* 11th edition. New York: The Population Council.

Omara-Otunnu, A. 1987. *Politics and the Military in Uganda 1890–1985*. London: Macmillan.

Population Reference Bureau (PRB). 1988. *World Population Data Sheet.* Washington, DC: PRB.

Romaniuk, A. 1980. "Increase in natural fertility during the early stages of modernization: Evidence from an African case study, Zaire," *Population Studies* 34, 2: 293–310.

Study Commission. 1986. Report of the Study Commission of United States Policy toward Southern Africa, in *South Africa: Time Running Out.* Berkeley and Los Angeles: University of California Press.

Teitelbaum, Michael S. 1992–93. "The population threat," *Foreign Affairs* 71, 5: 63–78.

United Nations. 1987. *World Population Policies,* Population Studies No. 102. New York: UN Department of International Economic and Social Affairs.

———. 1988a. *Statistical Yearbook 1985/86.* New York: United Nations.

———. 1988b. *1986 Demographic Yearbook*. New York: United Nations.

———. 1989. *South African Destabilization: The Economic Cost of Frontline Resistance to Apartheid.* New York: United Nations.

———. 1990. "Africa set for record population growth," *Africa Recovery* 4, 1: 36.

UNESCO (United Nations Educational, Scientific and Cultural Organization). 1989. *UNESCO Statistical Yearbook.* Paris: UNESCO.

UNFPA (United Nations Population Fund). 1992. *State of the World Population Report, 1991.* New York: UNFPA.

United States. 1989. *Statistical Abstract of the United States: 1989*. Washington, DC: Government Printing Office.

Valentine, Carol H., and Joanne E. Revson. 1979. "Cultural traditions, social change and fertility in sub-Saharan Africa," *The Journal of Modern African Studies* 17, 3: 451–472.

Wai, Dunstan M. 1981. *The African-Arab Conflict in the Sudan.* New York: Africana Publishing.

Wilks, Yorick. 1970. "Family planning, or tribal planning?" *Cambridge Review: A Journal of University Life and Thought* 92, no. 2198: 8–10.

World Bank. 1989. *Sub-Saharan Africa: From Crisis to Sustainable Growth, A Long-Term Perspective Study*. Washington, DC: World Bank.

# Fertility Control and Politics in India

V. A. PAI PANANDIKER
P. K. UMASHANKAR

INDIA WAS ONE OF THE FIRST COUNTRIES to introduce a policy intended to reduce the rate of population growth, yet to this day relative population size and fertility regulation remain contentious issues in the country's electoral politics. The interplay between attitudes toward population growth and electoral politics largely explains the marked differences among the states in the way the national family planning program is implemented. Before discussing India's current population policy and its implementation, a brief review of the history of the government's population control program is useful.

India made a modest beginning with a fertility control program in the first Five Year Plan commencing in 1951. Over the following two decades, the program was continued in the subsequent development plans with larger provisions and a wider reach. However, the program reached a crisis point in 1971. The findings of that year's census took the country's planners and policymakers by surprise: India's population growth rate had risen to 2.25 percent per annum. The "elite," which included politicians, planners, administrators, industrialists, and national media representatives, drawn mainly from the upper strata of society, sounded the alarm. The implications of unplanned population growth for planned development were discussed with great concern, and in October 1975 the Union Health Minister, Karan Singh, sent a note to Prime Minister Indira Gandhi stating that "The problem [of rapid population growth] is now so serious that there seems to be no alternative to thinking in terms of introduction of some elements of compulsion in the larger national interest" (Shah Commission of Inquiry, 1978: 153).

At the time that concern about this perceived population explosion was growing, the leaders of the central government, particularly the prime minister, were also involved in a political upheaval. In June 1975 the

Allahabad High Court nullified Gandhi's reelection to the Rae Bareli seat in parliament.[1] In order to protect her position of power, Gandhi moved swiftly to assert her authority by declaring an Emergency on 26 June 1975, suppressing all political dissent and imposing censorship on all media. Shortly afterward, Gandhi and the ruling Congress Party announced a Twenty Point Program to improve administrative efficiency and to ameliorate the condition of poor.[2] While the purpose of the Twenty Point Program was to centralize authority in the prime minister's hands, she justified it as promoting development in India.

During the Emergency, Sanjay Gandhi, the prime minister's younger son, assumed the leadership of numerous programs, including fertility regulation, adult education, and afforestation. Sanjay Gandhi held no official position in the government. He was new both to politics and to administration and underestimated the complexity of both; nevertheless, he exercised considerable influence on his mother and, through her, on the Congress Party.

The anticipated impact of unplanned population growth on planned development figured prominently in the new Twenty Point Program, and Sanjay Gandhi took it upon himself to promote population control. At that time, however, the states had sole authority to promote family planning and were reluctant to move vigorously in this area. Recognizing this reluctance, the Lok Sabha (the lower house of the Indian parliament), under the direction of Sanjay Gandhi and his mother, amended the Constitution in 1976 to give the central government greater powers to influence the implementation of the family planning program (Constitution Act, 1976). In January 1976, the prime minister announced that "We must now act decisively, and bring down the birth rate speedily too. We should not hesitate to take steps which might be described as drastic" (Shah Commission, 1978: 154). Following this announcement, the central government promulgated a new family planning policy in April 1976 that proposed bold and drastic measures to contain population growth (National Population Policy, 1976). The swiftness and decisiveness of the policy choices, made under a combination of political and population pressures, sent a clear message to party echelons, the administration, and the people that the government was resolute about the reduction of population growth.

Receiving these clear directives, the administration in the center and the states moved quickly, and much of the political leadership joined the drive to promote fertility reduction. Historically, sterilization targets given by the central government to state and local officials have been India's main approach to reduce fertility. Before the Emergency, these targets were neither reached nor enforced. However, during the newfound administrative push of the Emergency, local, state, and central government leaders zealously enforced and met these sterilization targets on a scale that brought about a tremendous popular reaction. Lest there be any doubt about what happened, the Shah Commission, which was later set up to investigate the excesses of the Emergency in Uttar Pradesh, may be quoted. According to the Commission's Report, in Uttar Pradesh even the police were involved in meeting sterilization targets:

> A Government order was issued on June 16, 1976 wherein, while fixing targets for various departments of the State Government, a target of 5,500 was assigned to the Police and Jail Departments. Subsequently . . . employees of the Police Department and the PAC [Provincial Armed Constabulary] were to achieve 7,500 cases and the targets for employees of the Jail Department were separate. (Shah Commission, 1978: 195)

As a consequence of the political-administrative push, the number of sterilizations rose from 1.3 million in 1974–75 to 2.6 million in 1975–76 and shot up to 8.1 million in 1976–77 (Shah Commission, 1978), a level it has not reached since. Uttar Pradesh, which could not achieve its target of 175,000 sterilizations in 1975–76, achieved 837,000 the very next year (ibid.). The government focused almost entirely on use of sterilization to reduce fertility; IUD and condom use showed little increase.

The stifling of political dissent, severe censorship, and overzealous implementation of sterilization targets led to strong public discontent, some of it expressed in violent resistance. In Uttar Pradesh alone, more than 1,500 people were arrested under a variety of existing statutes (ibid.).[3] Public anger was widespread and threatened the nation's well-being. Reading the signals, perhaps, Indira Gandhi announced a general election in January 1977 to take place the following March. In that election, Mrs. Gandhi and the Congress Party were swept away in the central government and in many state governments, including Uttar Pradesh and other states where the family planning program had been implemented with brutality.

The Janata Party, which won an overwhelming majority, moved into power in New Delhi and promptly disavowed Gandhi's fertility regulation program. The new government quickly announced a new population policy under a different name—the Family Welfare Program (India, Ministry of Health and Family Welfare, 1977). The message of this policy was clear. The program was to be voluntary. In response to the new voluntarism, the number of sterilizations dropped to barely one million in 1977–78 and other parts of the program also slowed down. The number of couples effectively protected by contraception, which rose from 18 million to 26 million between 1976 and 1977 (a 44 percent increase) at the height of the Emergency, dropped back down to 25 million (India, Ministry of Health and Family Welfare, 1992). The birth rate, which had fallen from 35.2 per 1,000 population in 1975 to 34.4 in 1976 and 33.0 in 1977, rose again to 33.3 where it remained until 1985 (Ministry of Health and Family Welfare, 1989). Family planning and population control became anathema to political parties and leaders alike. Though Indira Gandhi regained power in 1980, she and her party were very cautious. They had lost their enthusiasm for the family planning program.

The 1981 census gave India another jolt. It showed that the population continued to rise at an annual rate of 2.2 percent. Once again the political elite, including planners, administrators, industrialists, intelligentsia, and media, sounded the alarm. Government moved again, but haltingly. This time it was mute and cautious. The elite kept raising the issue but the political system did not respond. Even the media were subdued. Yet, as in his inaugural speech at the IUSSP General Conference in New Delhi in September 1989, Prime Minister Rajiv Gandhi remarked on "the cruel paradox . . . that in the ten-year period between 1971–81, the growth of population in India was the highest ever recorded in the history of the country."

During the 1980s, the Congress Party announced a health policy and promised a new population policy, but the latter was never formulated and nothing more was done except to allow the bureaucracy to tighten the implementation of the existing program. Each regime since has followed a policy of similarly cautious support for voluntary family planning. To conclude that the policy and program have collapsed would be wrong. Yet the conclusion is inescapable that, especially since 1976–77, the fertility regulation program in India has been a victim of electoral politics. The question is why?

The central explanatory propositions of this article are that India's policy of fertility regulation and its success or failure have been deeply conditioned by two central realities of the Indian polity: (1) India's diversity and (2) India's federal democratic political system. The complex interactive relationship between these key dimensions of the Indian political organization has determined the country's performance not only in the area of family planning but in other areas as well. The nature of these internal political factors, which have determined the electoral politics of fertility control to date and are likely to determine future performance, deserves further delineation. The changes wrought by India's development policies, including increased industrialization and economic growth since the 1950s, are the only antidote to the influence of population policy on electoral politics.

## India's diversity

Religious, ethnic, linguistic, and economic diversity characterizes Indian society. Of the estimated Indian population of 822 million in 1990 (Center for Monitoring Indian Economy, 1990), the religious composition based on the projection of the 1991 census is shown in Table 1.

The complexity of ethnicities, races, and castes is aptly illustrated by the estimate that Hindus alone have about 2,378 main castes, subcastes, or jatis. India's Backward Classes Commission estimated in 1980 that there

**TABLE 1 Percentage distribution of religions and number of adherents: India, 1991**[a]

| Religion | Percent of total | (N, in millions) |
|---|---|---|
| Hindu | 82.7 | (693) |
| Muslim | 11.3 | (94) |
| Christian | 2.4 | (20) |
| Sikh | 2.0 | (17) |
| Buddhist | 0.7 | (6) |
| Jain | 0.5 | (4) |
| Others | 0.4 | (4) |
| Total | 100.0 | (838) |

[a]Information on number of adherents and percent of total is not yet available from the 1991 census; the percent distribution from the 1981 census was applied to the population enumerated in 1991 to derive numbers and percentages.

SOURCE: Census of India, 1981, 1991.

are about 3,743 backward castes in India. Similar diversity prevails among the scheduled castes. The 1981 Census of India lists 1,086 scheduled castes, constituting nearly 16 percent of the population. The tribes in India number 566 (India's Backward Classes Commission, 1980).

Imposed upon the diverse caste groups is a multiplicity of languages, 16 of which, including English, have been officially listed in the Constitution. The 15 official Indian languages reflect the diversity of the federal structure, for in most cases one language covers a state. While Hindi is the official language of several states, some states recognize more than one official language. These languages all have rich traditions, literatures, and histories, and most enjoy distinct alphabets. A testimony to the linguistic diversity of India lies in the 1,652 spoken languages counted by the Census of India in 1961 (India, Registrar General, 1971). In subsequent censuses, the government classified many of these languages under 15 constitutional languages and 91 other languages (India, Registrar General, 1987).

The main problems that arise out of this linguistic diversity are how to provide education in the scheduled and other local languages at the primary level, and how to develop a national consensus on a language for official communication. Linguistic diversity also provides a challenging and baffling task in the field of information, education, and communication pertaining to fertility regulation and family planning.

In addition to religious, ethnic, and linguistic differences, India's great economic diversity, both vertical and horizontal, has been a source of constant debate and controversy within the country. Two central issues in this debate concern the problems of poverty and the large regional disparities in income. According to the 1989 *World Development Report*, the richest 20 percent of India's population account for 50 percent of all household expenditures, while the poorest 20 percent account for a mere 7 percent of such expenditures (World Bank, 1989). As of 1989 about 29 percent of the population lived below the poverty line, defined in India as an average consumption of 2,250 calories per day (India, Ministry of Health and Family Welfare, 1992). Within this aggregate picture, the poverty figure varies greatly from state to state.

One issue dominates India's economic, political, and social policymaking landscape: poverty. Because of the nature of the Indian political system, political leadership cannot ignore the pressure of some 300 million poor. The presence of this vast mass of poor people determines, to a large extent, public policy in the fields of development strategy, the system of political representation, and the family planning program.

In economic terms, three different Indias have emerged during the past several decades. First, the India of the reasonably affluent middle and upper-middle classes, which constitute about 25 percent of the population, or about 200 million. The second India, of about 40 percent or 330 million, has received some of the benefits of growth, both economic and political, since 1947 but manages a level of living that is not quite comfortable and yet not desperate. And the third India, which constitutes the bottom 35 percent, or 258 million, still lives in poverty.

Because disparities in education and literacy compound economic disparities, the major states have problems in accelerating social development, problems that are both social and economic in character. Education on the social side, and nonagricultural employment on the economic side, have clear impacts on fertility. Yet in the backward states of India, investments in both the social and economic sectors are still low. These various disparities pose serious problems for the Indian polity in general and for population policy in particular.

## India's federal democratic political system

Indian political organization is largely decentralized and works essentially through the elected government at the center in New Delhi and in the 25 states and 5 Union Territories. Soon after Independence, the Constitution, in force from 26 January 1950, distributed the powers of the Indian federation into three areas: the powers of the central government, the powers of the states, and the areas where power is shared but where central legislation takes precedence over state legislation in case of conflict.

Recurrent problems have existed in the relations between the center and the states since 1967, when the predominance of the Congress Party was first broken. In the parliamentary elections in November 1989, the Congress Party lost power for the second time at the federal level. As of October 1990, non-Congress Parties control 13 major states, ranging from Marxist governments in West Bengal and Kerala to centrist national parties like the Janata Dal in Uttar Pradesh, the largest state in India, and regional parties like the DMK in Tamil Nadu.

The multiparty political system of India represents one of the most complex democratic polities in the world. In the 1989 parliamentary elections, an electorate of more than 498 million cast votes in over 589,000 polling booths. The number of candidates who contested the 529 Lok Sabha (Lower House) seats for election exceeded 6,000; 110 political parties participated in the elections (Election Commission of India, 1990).

In addition to their political diversity, the states vary greatly in population size. Uttar Pradesh with an estimated population of 138 million in 1991 is larger than the majority of member countries of the United Nations, as well as larger than the population of every other South Asian country, including Pakistan and Bangladesh. Many other Indian States are large by comparison with most countries around the world. The population of Bihar, for example, was estimated in 1991 at 86.3 million, and Maharashtra at 78.7 million; Andhra Pradesh with 66.3 million, Madhya Pradesh with 61.1 million, Tamil Nadu with 55.6 million, and even Kerala with 29.0 million exceed many other countries in size (Census of India, 1991).

India has maintained its democratic traditions in spite of the gravest of provocations such as the recent powerful secessionist movements in Punjab and the Kashmir Valley. National elections have been held at regular intervals and have served to reinforce the democratic structure. Through the Panchayat system,[4] the democratic structure has also permeated the district, subdistrict, and village levels, and has been responsible for absorbing much social and political dissent and dissatisfaction.

The stresses and strains that this diverse democratic polity place on India should not be underestimated. In the early years of the Republic, linguistic riots virtually tore the country apart. In the 1980s and early 1990s religious fundamentalism such as that in Punjab has raised many fears. Some of the ethnic minorities in northeastern India have engaged in intermittent insurgencies beginning in the mid-1950s. Each one of these strains has forced the Indian polity to respond. Nevertheless, accommodation and representative government have been the two pillars of the Indian democratic solution to the problems arising from diversity. They have remained standing for more than five thousand years; their resilience and vigor have allowed the government to vary its response to population policy, based upon the country's social and economic diversity.

## Religion and the politics of fertility

Although Hindus constitute the overwhelming majority in India, the governmental system is far from monolithic. Many of the administrative districts upon which Lok Sabha electoral seats are based have a non-Hindu majority. For example, the majority in 12 of these districts is Muslim; in 2, Buddhist; in 14, Christian; and in 9, Sikh. In all, Hindus are in a minority in 56 districts, while at the state level, Punjab, Jammu and Kashmir, Meghalaya, Nagaland, Mizoram, and Arunachal Pradesh all have non-Hindu majorities.

In many of these states, some political elites, including Muslim and Sikh religious leaders, some political leaders, and even some political parties with a Hindu religious orientation, have had no inhibitions about promoting communal over national interests to gain political advantage. In general, religious leaders do not support the family planning program. Some of them, sensitive to the impact of a growing population on the quality of life of the people, remain neutral; others, confronted by measures intended to weaken the influence of religion upon the masses, have a tendency to use the fertility regulation program to provoke their followers and strengthen their political positions. In a newspaper interview in 1989, G. M. Shah, former Muslim Chief Minister of Jammu and Kashmir State, underscored the importance of religion for fertility politics in India, particularly in highly polarized non-Hindu states (Shah, 1989). In Shah's words,

> The government has hatched a conspiracy to reduce the Kashmiri Muslim population. Farooq Abdullah [then Chief Minister of the Jammu and Kashmir State] is an instrument of this plot. Our State had an 82 percent Muslim population in 1947; it is now a mere 54 percent as the 1981 census figures reveal. We should reject the government's family planning program. This is aimed at further reducing the Muslim population in Kashmir. Every Kashmiri Muslim should have four wives to produce at least one dozen children. (Shah, 1989: 33)

In fact, Shah's provocative arguments are spurious and his statistics are in error, for the Muslim population of Jammu and Kashmir State at the time of accession in 1947 was 57 percent and had risen to 64 percent by the time of the 1981 census. However, Shah's erroneous statistics are less significant than the thrust of his argument against family planning among Muslims. The Muslim population in Jammu and Kashmir is not the only group to point the finger of discrimination; the Hindu minority accuses the Muslim majority of systematic discrimination against Hindus, as do the Buddhists who are a majority in the Ladahk Division of Jammu and Kashmir State. This demographic rivalry dominates politics and society in Jammu and Kashmir. Wherever the population issue is sensitive, whether in Jammu and Kashmir where the Hindus are in a minority, or in other parts of the country where the population balance between Hindus and Sikhs or Hindus and Muslims is delicate, the theme is the same: Electoral groups perceive fertility control as reducing the political status and leverage of the majority community. Given the electoral arithmetic in more than 100, or almost 20 percent, of parliamentary constituencies, the family planning program is an exceedingly touchy political issue.

At the same time, the data also suggest that wherever educational levels and income per capita are high, the issue does not evoke the same popular response. For example, despite having a large Muslim and Christian population, Kerala, where social development—especially in terms of women's education—is high, the fertility of Muslims and Christians is not markedly different from that of Hindus. Moreover, the perception of a religious split in this southern state is markedly less. The situation is similar in Goa, where Catholics constitute more than 25 percent of the population.

In 1978, the Registrar General of India conducted a study of fertility by religious groups and found that, while the rural crude birth rate for Hindus was 32.6, it was 34.9 for Muslims, 25.7 for Christians, and 29.6 for Sikhs (India, Registrar General, 1979). Subsequent studies have also shown that birth rates among the different religious groups do not vary greatly. The same would hold true for caste and community structures. In fact, socioeconomic development softens the impacts of religious and ethnic identities on reproductive behavior. In other words, while religion and religious propaganda are important issues, the evidence shows that their impact on reproductive behavior in India depends a great deal upon the social development of particular regions of the country. Where educational levels, especially of women, are high, religion and religious propaganda do not have much impact on the adoption of family planning.

## Ethnic and caste diversity and fertility control

Ethnic and caste diversity also impinge on the acceptability and implementation of the family planning program. The backward castes—Hindus and non-Hindus—constitute more than 50 percent of the population. The intermediate castes worry about what they perceive to be a loss of their political power as a result of their relative decline in numbers. The problem becomes particularly acute when the political and economic interests of these intermediate castes come into conflict with those of the more backward castes or the scheduled castes.[5]

In recent years, political consciousness has been growing among the backward castes. In terms of numbers, they can control voting patterns and have the ability to influence the electorates' choices. In a situation of an increasing demand for affirmative action to improve access to education and employment for the backward sections, political influence is of critical importance. The backward classes, in their vast numbers, are entering onto the political stage and, more important, into the power structure. A growing proportion of ministers and chief ministers are drawn from their ranks.

For instance, after the February 1989 elections, both Uttar Pradesh and Bihar had chief ministers drawn from the Yadev, a backward caste. The backward castes, as a group, have observed the political power of numbers and are suspicious of proposals for controlling population growth. Brought up in a rural agricultural milieu where large families are respected, and where at least two sons are desired, members of these castes are highly suspicious of the perceptions of urban elites drawn from socially and economically advanced sectors of the population. One of the chief ministers of a state from a backward caste has nine children, and is reported to have said that he had this large family because he opposed the family planning program of the Congress Government.

Such distinct differences in perception make implementation of the fertility regulation program difficult. At the same time, there is growing appreciation by those leaders holding elected office that population growth is a problem. Many chief ministers and cabinet ministers of states appreciate the implications of population for their states' economic and social development. Lacking the confidence and ability to influence their electorates or even their parties, however, they often cede responsibility to the bureaucracy for administration of the program. This devolution of responsibility creates problems of its own; the programs tend to become rigid, poor in quality, and ineffective. Moreover, family planning is frequently resisted by the people it is designed to reach, a situation that makes program leaders feel politically insecure.

However, the spread of education in the intermediate castes and increasingly in the backward castes tends to reduce hostility to fertility control and the family planning program. Unfortunately, social development, including literacy, in the four largest states of India, Uttar Pradesh, Bihar, Madhya Pradesh, and Rajasthan, is moving at a very slow pace. The expenditure per capita on education in Bihar in 1986–87 was only Rs 63; it was Rs 72 in Uttar Pradesh, Rs 87 in Madhya Pradesh, and Rs 103 in Rajasthan. By contrast, the expenditure per capita in 1986–87 was Rs 150 in Punjab, Rs 180 in Kerala, Rs 200 in Delhi, and Rs 274 in Goa (India, Ministry of Human Resource Development, Education Department, 1989). These disparities in expenditure are indicative of the problem as well as the direction of the solution; they are evidence that most backward states have neither the financial nor the organizational resources to implement family planning policy.

## Economic diversity and fertility control

Of the three Indias referred to earlier, the first, the socioeconomic elite (the top 25 percent), continues to dominate the fields of administration, education, medicine, industry, and the media. Although they are gradually losing their direct hold on the political power structure, the upper classes are still in a position to influence the political system. This group practices family planning and is the main source of support for the program. They fear threats to their security and well-being from the burgeoning population.

The 35 percent or so of Indians who constitute the poor do not share the perception that population growth is an important issue. Those at submarginal existence levels regard large families as advantageous. An older child can look after a younger one, another can tend cattle, a third can gather firewood or help in domestic chores. The need to limit family size to improve one's own lot and that of one's children is not appreciated in a situation where sheer survival is the main concern. The middle 40 percent of the population are caught in between. The better educated among them share some of the attitudes of the elite and the less educated tend to retain the perspective of the poor.

Where infant mortality is high, having many children seems logical. Social customs, religious practices, and the fundamental realities of life cause people to favor large families. Women may want relief from repeated childbearing, but they lack the necessary status to have a say in reproductive decisions. Furthermore, many of them look with disfavor on the unimaginative and unsympathetic family planning program often presented insensitively by the administration.

The structure of employment, a second important dimension of economic disparity, is also of crucial significance for reproductive behavior. The available data indicate that for India as a whole, approximately 70 percent of employment is in the agricultural sector, some 13 percent in the industrial sector, and the balance is in the service sector (World Bank, 1988). In such backward states as Uttar Pradesh, Bihar, and Madhya Pradesh, agricultural employment still stands at 70 percent, whereas in the more advanced states of Maharashtra, West Bengal, Tamil Nadu, and Kerala it has dropped to about 50 percent. The economies of these latter states show a rapid growth in nonagricultural employment, especially in the service sectors like transport, communications, and banking.

Disparities in education compound the impact of economic inequality on fertility control. Education has also had an impact on health and social factors that promote fertility regulation. In 1988, Kerala, with 65.7 percent of its female population literate and 100 percent enrollment of girls in middle schools, had an estimated crude birth rate (CBR) of 19.9, a crude death rate (CDR) of 6.3, and an infant mortality rate (IMR) of 28 per 1,000 live births. At the other end of the scale, Rajasthan, with a literate female population of 11.4 percent and with an enrollment of girls in middle schools of only 16 percent, had a CBR of 32.8, a CDR of 13.2, and an IMR of 103. The mean age of marriage for girls in Rajasthan is 16 years, and the percentage of women employed in the organized sector is only 11.4 percent (Ministry of Health and Family Welfare, 1989; India, Registrar General, 1989; Sharma and Retherford, 1987). The central government, which shares the responsibility for economic and social planning, including population control, appreciates that effective implementation of the family planning program would require considerable efforts by these states to improve the level of social and economic development. Unfortunately, these efforts are not forthcoming in adequate measure.

## The federal polity and problems of fertility control

When the Constitution was amended in 1976, the central government placed population control and family planning on the "Concurrent List," which confers power on the central government to pass legislation in the event of a conflict between the states and the center. However, since the Government of India has no direct machinery for implementing the Family Welfare Program, it has to depend upon the machinery of the states for carrying out the national program of family planning.

The primary instruments available to the Government of India in this endeavor are the national development plan that the government formulates every five years and the allocation of resources for fertility regulation under the plan. In 1991–92, the outlay on the program was scheduled at approximately Rs 7.49 billion (US$242 million) (India, Ministry of Health and Family Welfare, 1992). The states do not generally invest any of their own resources in the program, although a few add some incentives and disincentives, financial and nonfinancial, to make family planning more attractive to residents. The reality is, however, that the fertility control program has hardly any political support in most of the states, and if it were taken off the centrally sponsored list, most states would likely opt out of the present scale of the program. The nature of the party rule in the states further compounds the problem. The state governments have their own priorities arising out of the political orientation of the party in power. Clearly, some do not wish to be involved with the family planning program.

To the credit of the Indian government, however, there is an unusual consensus among the principal national political parties on the need for fertility control to reduce the pressure of population on the nation's financial and physical resources for development. All three of the major national political parties, the Congress Party, the National Front, and the Bharatiya Janata Party, found a place for fertility control in their 1989 election manifestos. Few national leaders from these parties oppose the family planning program. The National Development Council, which is currently the highest forum for discussing the national development agenda, and which consists of the prime minister, his cabinet colleagues at the center, and all the chief ministers, has never witnessed any objections to the fertility control program. Nevertheless, while the National Development Council and the five-year plans have always taken note of the need to manage population growth and proposed appropriate schemes, population control has never emerged as a priority for action in either the council's deliberations or in the five-year plans. This is largely a consequence of the political sensitivity of the program especially after the excesses of the Emergency during 1975–77.

A number of states, particularly those like Tamil Nadu where regional parties are in power, are apprehensive about pursuing their moderately successful family planning programs. They express fear of the burgeoning populations of the larger and more backward states. Tamil Nadu and similar states worry that the size of the populations in the backward states may become the basis for establishing political dominance and for determining the sharing of national resources. They would prefer other criteria for the allocation of resources that are not based directly on population size. Some would even like to institute disincentives for those states that do not pursue an effective population control program.

Below the national level, the political parties display a variety of differing attitudes toward fertility regulation. Some regional parties, or parties that dominate a state, favor population control as they observe that increasing population is harming the development of the state. As these parties have no interests outside their own state, they do not perceive lower fertility as a threat to their hegemony. Some such parties, indeed, actively promote family planning. The leftist or Marxist parties tend to subscribe to the fertility control program, and while they do not openly promote the program, their cadres lend support at the local level where it counts most. The other national parties, ranging from conservative to socialist, while subscribing to the concepts of family planning, hold back open support for fear of offending their electorates. All the national political parties, having seen the effects of the 1975–77 campaign, are wary of espousing fertility regulation lest their electoral interests should suffer. As of 1990, for most states, and certainly for the more backward states that account for much of the population growth in India, the family planning program is not a political program of high priority. Given a choice, most leaders of state governments would divert the central funds earmarked for fertility control to other programs that they feel have greater importance and acceptability among their constituencies.

This reality of the government structure in India remains a hindrance for the national population policy because, ultimately, even if the family planning program is fully funded by the central government, implementation of the program lies entirely with the state governments. This division of responsibility means that the center only proposes and the state disposes, depending upon its perceived self-interest, the quality of its political leadership, its level of commitment, and its administrative competence. These factors are, therefore, reflected in the performance of the program in different states, among which the disparities are striking.

The dilemma of the central government is a genuine one. On the plane of electoral politics, it has to reckon with the attitudes of the states and also its own survival. But on the development side, the government has to ensure a return for the resources deployed and also tackle a problem that impedes the attainment of higher standards of living for the Indian people.

## India's democratic political system and fertility control

The nature of India's democratic system greatly compounds the pressures and pulls upon the federal polity. In the initial phase of the Indian democratic evolution, from 1947 to about 1967, a single party, the Congress Party, dominated, and it was led by a relatively small, Western-educated urban elite. The picture changed rapidly after the parliamentary elections of 1967 when, for the first time, the Congress Party was routed in many northern states. The picture has continued to vary ever since. Single-party dominance in India is over, and the democratic structure is dominated today by parties with widely differing political ideologies. Political power in India has been transferred from the English-educated elite to the elites educated in regional languages in almost all the major states.

This transfer of political power is one of the most important achievements of Indian democracy. Yet, the very transfer of political power to the regional elites brings in its wake the political compulsions of emerging local power groups. The aspirations of these groups have a major bearing on their attitudes toward fertility control programs. For instance, several intermediate and backward castes that are coming to power, especially in the large northern states, are unenthusiastic about the family planning program. In addition, neither the Muslim minority nor its leadership is supportive of family planning.

Given the balance between the various castes among the Hindus and minority groups,[6] political compulsions and electoral politics militate against active support of the fertility control program. No caste or community is willing to support the population program vigorously at the grass-roots level. In India, a broad national consensus on the importance of the family planning program confronts an apathetic, at times hostile, political system at the local level where implementation must take place.

Since the 1950s, the motivating force for the adoption of fertility regulation and support for the national program has been economic and social change, especially in education and health. Political support for the family planning program in India remains in large measure confined to members of the educated, nonagricultural elite. Where social and economic development has proceeded sufficiently, caste and religious affiliations no longer seem to determine attitudes toward the population program. Members of this constituency, comprising perhaps 40 percent of the Indian population, have already adopted a two-child norm. For the remaining 60 percent who still conform to a two-son norm, the electoral process continues to create impediments to the acceptance of the population control program.

## Conclusion

In agricultural and traditional societies, fertility control has historically been difficult to introduce. In India, the family planning program is at the heart of the politics of an immensely diverse federal and democratic structure. Diversity brings in its wake the enormously difficult problem of identity. Each identity-bound group, whether religious or ethnic, caste or linguistic, feels that its safety lies in numbers and percentages; the moment its demographic leverage decreases, its importance and influence in government suffer, and its vital interests are at stake.

This anxiety over numbers is the central problem of the Indian family planning program. Developmental politics is the only way to solve the problem. Eventually, as economic and social development broadens and accelerates, developmental politics will probably win. But, for the time being, electoral politics will dominate the family planning program of India, making its progress slow and painful.

## Notes

1 Rae Bareli was Indira Gandhi's parliamentary constituency. The Allahabad High Court convicted Mrs. Gandhi on a charge of electoral fraud and disqualified her from holding office for a period of six years.

2 Numerous analyses of the Emergency have been put forth. For two that deal specifically with the family planning program, see Pai Panadiker et al., 1978 and Gwatkin, 1979.

3 According to the Shah Commission (1978), the breakdown of arrests is as follows: 62 under the Maintenance of Internal Security Act; 1,159 under the Defence of India Rules; 303 under the Indian Penal Code; and 20 under the Criminal Procedure Code.

4 The Panchayat system consists of directly and indirectly elected bodies at the village, subdistrict, and district levels. Set in place in the late 1950s and early 1960s, the Panchayat system has revenue-raising power and is responsible for planning and assisting with the implementation of all development projects, including those of the central and state governments.

5 Scheduled castes, also called untouchables, are a group of lower castes in Hindu society. India's constitution entitles these groups to certain privileges and concessions. Backward castes are a cluster of educationally and socially disadvantaged castes belonging to all religious groups.

6 This designation refers to certain religious minorities, such as Christians, Muslims, and Sikhs.

## References


Census of India. 1981. *Provisional Population Tables.* New Delhi.

———. 1991. *Provisional Population Tables.* New Delhi.

Center for Monitoring Indian Economy. 1990. *Basic Statistics Relating to the Indian Economy,* vols. 1 and 2: All India. Bombay.

Election Commission of India. 1990. *Report of the General Elections to the House of People and Legislative Assemblies, 1989–90.* New Delhi.

Gwatkin, Davidson R. 1979. "Political will and family planning: The implications of India's Emergency experience," *Population and Development Review* 5, 1: 29–59.

India, Ministry of Health and Family Welfare. 1977. *Family Welfare Program—Statement of Policy.* New Delhi.

———. 1989. *Family Welfare Program in India Yearbook, 1987–88.* New Delhi.

———. 1992. *Health Information of India 1992.* New Delhi.

India, Ministry of Human Resource Development, Education Department. 1989. *Selected Educational Statistics, 1987–88.* New Delhi.

India, Planning Commission. 1989. *Annual Plan, 1989–90.* New Delhi.

India, Registrar General. 1971. *Census of India 1971.* New Delhi.

———. 1979. *Level, Trends, and Differentials in Fertility.* New Delhi.

———. 1987. *Level, Trends, and Differentials in Fertility.* New Delhi.

———. 1989. *Level, Trends, and Differentials in Fertility.* New Delhi.

India's Backward Classes Commission. 1980. *Report.* Delhi: Controller of Publications.

*National Population Policy Statement by Dr. Karan Singh,* Minister of Health and Family Planning. 16 April 1976. New Delhi: Ministry of Health and Family Planning.

Pai Panadiker, V. A. et al. 1978. *Family Planning Under the Emergency: Political Implications of Incentives and Disincentives.* New Delhi: Radiant Publishers.

Shah Commission of Inquiry. 1978. *Third and Final Report.* New Delhi: Ministry of Home Affairs.

Shah, G. M. 1989. "Every Kashmiri Muslim should produce one dozen children," *The Illustrated Weekly of India,* 2 April: 33.

Sharma, O. P., and Robert D. Retherford. 1987. *Recent Literacy Trends in India.* New Delhi: Registrar General.

United States Department of Commerce. *Population Trends: India.* Washington, DC: Government Printing Office.

World Bank. 1988. *World Development Report 1988.* New York: Oxford University Press.

———. 1989. *World Development Report 1989.* New York: Oxford University Press.

# Demographic Dynamics and Development: The Role of Population Policy in Mexico

GUSTAVO CABRERA

IN THE LAST 50 YEARS, Mexico has experienced vast demographic changes that have had significant implications for its social, economic, and political life. Between 1940 and 1970 mortality declined sharply, resulting in one of the highest rates of population growth in the world. This period also saw the country's urbanization and the emergence of Mexico City as a metropolitan area. Toward the end of this period, in the late 1960s, a steep decline of fertility began. These three demographic changes have occurred within a relatively short time and have shaped the "population problems" that have figured in national debates, not only in the past 50 years, but throughout Mexico's history.

Since its independence in 1821, the Mexican state has periodically attempted to produce demographic change through laws and administrative actions. For more than a century and a half, population policy was profoundly pronatalist; since the mid-1970s, the Mexican government has implemented programs to limit population growth. This chapter focuses on the demographic changes experienced in Mexico during the past 50 years as outcomes of its population policies.

## Background: Population in Mexican political thought and law

### The nineteenth century

From its origin as a state, Mexico's political organization has confronted problems relating to its population, its territory, and its sovereignty. The size of the population, its distribution, and dynamics, in conjunction with the problem of assuring national security to a sparse population in a large territory, were considered in the Federal Constitution of the United States of Mexico in 1824 (Estados Unidos Mexicanos, 1989). With a land area of more than 4 million square kilometers (Tamayo, 1962), twice the size of Mexico today, and a population of only 6 million compared with more than 80 million today, the government experienced great difficulties in assuring national security and the welfare of the population (Loyo, 1935).

During the first decades after independence, colonization programs were established in which immigrants received special treatment. Laws and measures to promote immigration were put into effect, though not always with the desired results. Groups of foreigners arrived who did not integrate into Mexican society. Certain areas of the country, such as the northern border, were nearly ungovernable. In 1835, the population in Texas was estimated at approximately 35,000 free inhabitants of Anglo-Saxon extraction and 5,000 slaves, while the Mexican population in the state numbered fewer than 5,000 (Zorilla, 1977). This disparity is partly explained by the ample collaterals offered to settlers on an open border, the lack of a vision of Mexico's future development, and the important role assigned to foreign colonization.[1]

### The twentieth century: The Porfirio Díaz period and the revolution

During the first decade of the twentieth century, still under the presidency of Porfirio Díaz (1876–1911), Mexico continued its struggle to attract foreigners without noticeable results. The first coherent set of regulations for the movement of foreigners appeared in the 1908 Immigration Law. The country's social conditions improved as a result of the government's economic policy, which brought about a decline of the high mortality levels. Mexico's population increased from 13.6 million to 15.1 million, reflecting an annual growth rate of 1.1 percent, mainly as a result of natural increase. Over the decade, applications for admission to the country numbered about 300,000. However, 60 to 70 percent of the applicants were not admitted because they were sick or considered undesirable. The number of foreigners living in Mexico in 1910 was estimated at 116,400, which represented only 0.77 percent of the total population (González Navarro, 1974).

The outbreak of the Mexican Revolution in 1910 interrupted demographic expansion. During the ten years of armed conflict, the population decreased from approximately 15.1 million in 1910 to 14.3 million in 1921. The decrease was a result of the combination of deaths directly related to the armed struggle, those caused by the "Spanish influenza" epidemic, Mexican emigration to the United States, and the decline in the birth rate resulting from the temporary separation of married couples and the postponement of new unions. However, the decrease cannot be determined precisely because the results of the censuses of 1910 and 1921 are in doubt.[2]

## The post-revolutionary period

Once the revolution ended, the country's reconstruction began with a series of institutional and political changes designed to consolidate the government. Between 1921 and 1930, the population growth rate rose again to an annual 1.1 percent, yielding a total population of 16.5 million in 1930 (Dirección General de Estadística, 1934).

The governments in power after the revolution once again promoted international immigration and the settlement of the country. Mexico was flooded with European immigrants wishing to enter the United States. Little control existed over the foreigners who were already settled or those who had recently arrived. The government, therefore, enacted the Migration Law of 1926, which required the registration of foreigners and paid special attention to the tourist status with the objective of attracting permanent immigration. The immigration of workers was limited to those regions and occupations where there was a need for labor. One provision of the law forbade entrance to individuals engaged in illegal drug traffic. The law also regulated the ever-increasing emigration of Mexican workers to foreign countries, especially to the United States. In 1930, modifications of the 1926 law introduced precise conditions for immigration (Sierra Brabatta, 1988; Leal, 1975; González Navarro, 1960). The Advisory Council on Migration was established as an interministerial governmental agency for the implementation of this law. The council was the forerunner of the present National Population Council.

In spite of the continuing pronatalist stance of post-revolutionary governments, isolated attempts to promote contraception were made during these years. In 1916, the first Feminist Congress was held in the state of Yucatán, in southeastern Mexico, chaired by the state governor, a celebrated and respected agrarian leader. One outcome of this congress was the founding of the Feminist Council, which proposed women's emancipation. At the Regional Workers Convention held in 1917 in the city of Tampico, the right to prevent "unlimited procreation" when it affected workers' standard of living was recognized (González Navarro, 1974). In 1921, another Feminist Congress met in Mexico City to demand women's right to vote. In 1922, Margaret Sanger, the American women's rights activist, published a booklet, *The Home Compass: Safe and Scientific Contraceptive Methods,* that was widely disseminated throughout Mexico. (Reactions to the booklet were soon forthcoming. The Caballeros De Colón [Knights of Columbus], a Catholic lay group, addressed a letter to the authorities demanding the prosecution of the press for the crime of publishing Sanger's booklet.) In 1925, during the presidency of Plutarco Elias Calles, three clinics were opened to provide services to women who wished to regulate their fertility (De Miguel, 1983).

### The new nationalism

Legal regulations in population matters were still far from complete. Under President Lázaro Cárdenas, who took office in 1934, population policy was changed to conform to the new nationalism derived from the revolution. Immigration was no longer considered the way to settle the country. President Cárdenas's population policy (Primer Plan Sexenal 1934–1940) was calculated to enable Mexico to escape from its economic and social backwardness by focusing on specifically Mexican solutions.[3] Under these nationalist principles, Mexico's first General Population Law was elaborated and issued in August 1936, replacing the Migration Law of 1930. The 1936 law instituted a new interministerial agency, the Advisory Council on Population.

Under the 1936 law, demographic momentum was to be achieved through the encouragement of natural growth, the repatriation of nationals, and, to a smaller degree, immigration. Marriages and births were to be encouraged and the high incidence of general and infant mortality was to be lowered. To improve the distribution of the population, the government attempted to direct migration to various states and to promote ethnic intermarriage. An effort was made to increase *mestizaje*—mixed marriages between individuals of Spanish and Indian descent. According to the General Population Law, restrictions were placed on the emigration of nationals in order to avoid population decline.

The law required the establishment of differential quotas for foreign nationals who wished to live in Mexico. These quotas were enacted shortly afterwards as follows: For 1939, the first year of application, no limit was set on the number of individuals coming from countries of North and South America and Spain; as many as 1,000 immigrants were accepted from Western and Central European countries and Japan; and only 100 immigrants were permitted entry from all other countries combined. Thus, the national policy placed boundaries on the number and diversity of foreigners admitted to Mexico (De la Peña, 1950).

Health improvement programs became the key means for the state to implement population policy and reduce high mortality rates. Public health programs and supporting infrastructure were incorporated into the social and economic development strategies adopted by the Cárdenas government and those that followed. A Social Security Law was enacted, the Mexican Institute for Social Security created, and a series of health measures and campaigns put into effect, accompanied by substantial expenditures on health and other social programs including education.

The Mexican government took several measures to increase the country's fertility levels. The notion of Mexico's greatness and its relation to a large population were incorporated into school textbooks, and large families were officially encouraged by social recognition and monetary awards. These elements accorded with popular religious sentiment and national pride. The sale of contraceptives was forbidden. Some states in the federation imposed stringent measures. The government of the state of Tamaulipas in northeastern Mexico, for example, established a tax on unmarried people over 25 years of age, and on divorced or widowed people without children. The tax was to be paid according to a progressive schedule that ranged from 5 to 20 percent of wages (González Navarro, 1960).

By 1947, the population of Mexico was increasing at an annual rate of 2.7 percent. That year, President Miguel Alemán announced that the Mexican population was increasing by half a million each year, a rate that assured the country's economic development. Over time, the objectives of the pronatalist laws were attained; and between 1940 and 1970, as mortality declined, Mexico's great demographic expansion took place.

The expansion of these decades was attained primarily through natural growth. The size of the foreign population was insignificant; between 1930 and 1970, immigration accounted for fewer than 150,000 new residents per decade. The nationalities of the immigrants clearly reflect the application of the quota system mandated by the *Dirección General de Estadística* of 1930, 1940, 1950, 1960, and 1970. Thus, the government's earlier attempts to settle the country through encouraging immigration were reversed.

The demographic changes that took place during the period between 1934 and 1970 can be summarized as follows: (a) the size of the population and its rate of growth both trebled in only 40 years; (b) because fertility remained high while mortality declined, the age structure became younger, with nearly 50 percent of the population being younger than 15 years by the end of the period; (c) urbanization increased and Mexico City became a metropolitan area as a consequence of the intensified migration of peasant workers and families to large towns; (d) the emigration of Mexican workers to the United States increased; and (e) immigration continued to be an insignificant factor in Mexico's population growth.

The governments of this period seemed satisfied with the level and swiftness of population growth.[4] By 1970, Mexico had 50 million inhabitants.

## Toward a new population policy

Not until the early 1970s did explicit change occur in political thinking about population that led to adjustments in the legal framework for population policy. This new thinking grew out of changes in the political and social contexts that had been defined in the late 1950s and in the 1960s. Mexico engaged in a debate on the role of population in the development process, which later gave way to a more focused discussion of population issues. Opposing positions in the discussions reflected differing political and ideological leanings.

The population debate of the early 1970s took place against the background of favorable rates of economic growth, averaging between 6 and 7 percent per year (Nacional Financiera, 1978). These rates were twice the high population growth rates of 3 to 3.5 percent and created an artificial perception of congruency between rapid population growth and development.

The development policy that produced such rapid economic growth had modified the economic and occupational structures and enabled them to absorb both the high rates of natural increase and migration from rural to urban areas (Sandoval, 1988). The facts seemed to confirm the pronatalist optimism of successive Mexican governments. Nevertheless, some years earlier national and state political and governmental officials had begun to express concern about what they referred to as Mexico's "demographic problem."

In 1966, the government of Tabasco pointed out that the territory of the state was not expanding, while the number of inhabitants was increasing explosively. Within 10 to 15 years land availability would become a serious problem. The governments of the states of Jalisco (1956), Nuevo León (1966), and Tlaxcala and Veracruz (1967) declared themselves unable to satisfy the existing demand—to say nothing of future demand—for educational services because the rates of population growth outstripped available resources. Already in the late 1960s, the government of Nuevo León, the state with the best welfare conditions in the country, was openly pessimistic about its "threatening demography." The government argued that the population was growing at a rate no economic development could match and was, therefore, likely to lower the standard of living (González Navarro, 1974).

During this period, members of various political circles proposed solving the population's needs by means of social programs, the assignment of lands for the peasants, and the creation of a sufficient number of jobs. However, no one proposed the implementation of a new demographic policy to reduce high fertility and the rate of population growth. The states hoped to accommodate themselves to constant demographic growth by demanding more resources for social and economic programs. Although Mexico had a legal framework that empowered state governments to formulate laws, no government adopted its own population legislation.

### Early steps in family planning

Article 40 of the Political Constitution of the United States of Mexico determines the political organization of the Mexican Republic and allows each federal entity (state) to have its own political constitution. Consequently, state governments have the right to create their own laws so long as these laws are not in opposition to the country's political constitution or a federal law, as well as to organize their own development projects by means of social and economic policies.

In the matter of population policy none of the states exercised this right. They were prevented from doing so by the tradition of the Mexican political system, which can be defined as formally federalist but centralist in fact. Power is primarily concentrated in the nation's highest authority (Meyer, 1974), and the president is the only person who can enact national policies (Márquez, 1984). On such a complex political issue as population policy changes, with important and controversial ideological overtones, state governments would not be allowed, nor would they dare, to express themselves independently.[5]

The first steps toward the introduction of family planning services were taken by groups of interested citizens organized by professionals, academics, and entrepreneurs. Some groups were concerned with the consequences of early and frequent pregnancies, while others sought to decrease the apparently large number of illegal induced abortions. The most persuasive arguments used to effect change, however, focused on the need to reduce fertility. These arguments brought about the establishment of a number of private, nonprofit agencies dedicated to clinical research and the training of doctors and nurses in population issues and family planning. Among the agencies created at this time were the Association for Maternal Health, in 1959; the Center for Research on Fertility and Sterility, now renamed the Mexican Foundation for Family Planning, in 1964; and the Foundation for Population Studies, 1965 (Asociación Mexicana de Población, 1979; Sandoval, 1988). Financial support for these associations was provided by private organizations in the United States and international agencies, including the International Planned Parenthood Federation (IPPF), the Ford and Rockefeller Foundations, and the Population Council.

During the 1960s a number of academic institutions were established to study demographic trends, their causes and consequences, and likely future developments. Thus, in 1960, the Mexican Institute for Social Research was created, followed in 1961–62 by the Institute for Social Research at the Autonomous National University. Also important was the foundation in 1964 of the Center for Economic and Demographic Studies (CEED), currently the Center for Demographic and Urban Development Studies (CEDDU), at the College of Mexico. The CEED developed a broad program of population research and introduced the first Latin American master's degree program in demography. In 1973, as the new population policy was being introduced, the Mexican Population Association was formed.

Bringing together many nationally known individuals from business and the professions, these two groups of organizations, academic and professional, became the leading advocates of family planning and, in some cases, of a new population policy. During the 1960s, and even earlier, three positions expressed by these groups can be distinguished. One current of opinion opposed formal birth limitation programs (see Turner, 1974; Nagel, 1978). Individuals taking this position were not opposed to granting couples and individuals free choice in handling their fertility (Durán Ochoa, 1955; Loyo, 1960; Benítez Zenteno, 1967). They believed that if economic growth were assured, resources were redistributed, social development that included the whole population were sustained, all of which were fundamental government responsibilities, a decline in fertility would occur as a result, and specific fertility-limitation programs would not be a priority.

A second current of thought, shared mainly by members of medical groups affiliated with private agencies offering family planning services, endorsed family planning programs to reduce fertility rates.

The third position favored the creation of a new population policy that would include family planning programs as an important component. Individuals in this group argued that although political and economic transformations were the basis for improvements in welfare and social justice, an explicit population policy, drawn within the legal framework of liberties and human rights conferred by the country's political constitution, was necessary to encourage demographic change and hasten the attainment of balance between population and development (Urquidi, 1967, 1969, 1973; Cabrera, 1966; Lerner, 1967; Alba et al. , 1973; Navarrete, 1967).

The position taken by the Catholic Church came as a surprise to those engaged in the debate. The Church was expected to have great influence on the debate held in Mexico, influence to sway the government decision by opposing the creation of official family planning programs. In fact, the influence and participation of the Catholic Church had little bearing on events. Since the Mexican Revolution, the Church has had a symbolic and expressive function rather than a political one, and it has developed great tolerance regarding the contraceptive practices of its congregation (Márquez, 1984).

The institutionalization of sociodemographic research in various academic centers of the country also contributed to creating awareness among political elites of population problems and their effects on development (CEED, 1970). A group of academics proposed to elaborate a new General Population Law under which population policy would be integrated with development policies. The objective of the law would be to influence the size, growth, age structure, and distribution of the population. Family planning would be one component of the policy, and would be supported by education, social communication, and medical services offered by the public sector.

Throughout the 1960s and early 1970s, the debate on population and the emergence of new ideas and principles concerning the subject were accomplished without opposition to family planning. Instead, the need for family planning programs was accepted generally, provided that three conditions could be met. First, family planning programs were not to be considered as a substitute for development; second, individual and human rights were to be respected; and third, there was to be no foreign influence or pressure upon decisionmaking on this subject, so that national sovereignty would be preserved.

### The new presidential position

The results of the 1970 Population Census confirmed that Mexico's population had grown to more than 50 million, with an annual intercensal growth rate of 3.5 percent (Dirección General de Estadística, 1973). If this growth rate continued, Mexico would have approximately 135 million inhabitants by the year 2000. This rapid growth notwithstanding, during his political campaign as presidential candidate in 1969, Luis Echeverría reaffirmed the historical pronatalist thesis that "to govern is to populate" (Alverdi, quoted in Warwick, 1982: 19).

In his inaugural address as President of Mexico, on 1 December 1970, however, President Echeverría changed his mind, saying:

> Today Mexico faces a situation whose nature and magnitude were not foreseen early in this century. Its population has multiplied more than thrice since the end of the armed movement. . . . We must precisely define the model of country we want, and can be, by the turn of this century, giving way, from this moment on, to the qualitative changes required by our organization. (Echeverría, 1970)

With this introductory speech, the president indicated his intention to reform past laws and to take a new direction with regard to population. He was becoming aware of the implications of demographic trends for continued development. This change of heart bears the influence of a number of people who favored a new population policy. Among these leaders Víctor L. Urquidi, at that time president of the College of Mexico, was especially influential.

During the first three years of President Echeverría's government, the idea of a new General Population Law was elaborated, largely under the aegis of the Secretary of the Interior, Mario Moya Palencia, its main promoter. In his address to the Union Congress on 1 September 1973, President Echeverría observed:

> This is the moment to seriously consider a problem which, for a long time, has been faced by many nations with different economic and political struc-

> tures. Several sectors of our population face the problem of family growth. Mexican women attend health centers and official and private clinics by the thousands, demanding guidance on the possibility of fertility regulation. We reject the idea that a purely demographic criterion directed to reduce births can substitute for development's complex enterprise. But we would be badly mistaken if we were to disregard the seriousness of population increase and the needs it generates. (Echeverría, 1979)

The announcement of the government's new position on population paved the way for the elaboration of the new General Population Law, which was submitted to the Union Congress and approved in December 1973, taking effect in January 1974. The law provided for the establishment of a new National Population Council, a government agency created for the sole purpose of implementing the population policy. This limited mission represented a change in the council's responsibilities that, under the 1947 law, had had only a consultative status.

In summary, the legal reversal from pronatalism was brought about by the conjunction of a number of forces and processes, including the experience gained from family planning programs within the private sector. The proof that the Church was not an obstacle to family planning programs was also important (Episcopado Mexicano, 1973), as was the demonstration that some progressive Catholics accepted family planning and freedom of choice although they rejected abortion and any form of coercion. In addition, a number of influential political and academic personalities and social groups favored a new population policy. The credibility of academics was enhanced by progress in sociodemographic research, particularly by work on long-term demographic projections. This work made clear the pressures that would limit the government's capacity to carry out necessary economic and social programs if the rate of population growth were not curbed. A final factor was the presence in Mexico of such international agencies as the United Nations Fund for Population Activities and the World Health Organization, and many other North American and European donor groups, that were willing to provide technical assistance and economic support for the implementation of family planning programs.

The relative weight of the factors that contributed to the government's growing political will to carry out a new population policy is difficult to determine. However, the government's ability and sensitivity should be acknowledged, since its public statements and actions were based on the principles of different social sectors and institutions. Examples of the government's receptivity were its acceptance of the "responsible parenthood" concept (Episcopado Mexicano, quoted in Warwick, 1982), proposed by the Catholic Church, and of the need for state intervention in development planning, which is closely related to the demographic behavior of society. The preservation of national sovereignty, human rights, and the freedom of choice of couples and individuals were concepts that legitimized the new population policy.

### Bases and strategies for the reduction of demographic growth

In 1974, the Mexican political system moved to establish the legal context for a new population policy. The objectives of the policy were to reduce the rapid rate of population growth and to modify the population's age structure and its distribution throughout the territory. Simultaneously, the new policy was intended to spread a more realistic view of the country's economic capacity and natural resources and to arouse concern so as to reduce future demands on scarce resources.

The political, legal, and administrative framework for the national family planning program was established against the background of the General Population Law, the establishment of new political and bureaucratic agencies to implement it, and necessary amendments in the constitution.[6] The government sought to develop the program in an orderly fashion to facilitate subsequent monitoring and evaluation. Health and social security agencies immediately began to offer family planning services on a massive scale, and the mass media presented widespread programs on family planning. In order to coordinate the family planning programs, the Interinstitutional Commission for Maternal-Child Care and Family Planning of the Secretariat of Health (Secretaría de Salubridad y Asistencia) was created in 1977. The commission was intended to foster better communication among the governmental agencies that provide medical care. It laid the groundwork for the government's effort to include family planning services in the program of health policies.

In 1977, the National Population Council publicly stated the principles, objectives, and goals of the National Population Policy and the National Family Planning Program (Consejo Nacional de Población, 1980). Demographic growth targets were proposed. They included reductions in the rate of growth from 3.2 percent to 2.5 percent between 1976 and 1982; and to 1.9 percent by 1988, 1.3 percent by 1994, and 1.0 percent by the year 2000. If these targets were met, Mexico's population was projected to be approximately 100 million by the turn of the century, 32 million less than the population projected should the existing growth rate continue. The contraceptive prevalence targets required to achieve the policy objectives were set by the National Family Planning Program and constitute one of the most important aspects of the plan.

This new stage in the development of the population policy was achieved under the political conditions stated in the amendment to the General Population Law. The Federal Executive Power, through the president of the republic, took the initiative in reforming the constitution and the new law, and in submitting them to the Union Congress. Consequently, the way was paved for the free and sovereign state governments to participate in this new public policy. Thus, by 1985, each of the 31 states had created its own population council in the image of the National Population Council, with representation from their highest political authorities. State governments were accorded the responsibility of carrying out their own population policies, supported by the technical assistance of the National Population Council. In this way, locally based political and civil leaders became involved in the efforts of the Federal Executive Power to carry out the policy in a decentralized manner.

The current federal family planning program is required to assume a dual coordinating function. First, through the federal Secretariat of Health the program is required to coordinate with all federal health institutions and programs; second, the federal program is also required to coordinate with family planning programs administered by the health institutions of the individual states. This multiple coordination task is both logistically and politically complex. While the program has been generally successful, the achievement of demographic goals and contraceptive coverage differs among regions. The differentials are clearly related to the varying levels of socioeconomic development and urbanization among the states. States with more advantageous social conditions, greater governmental awareness, and more active participation of local institutions have achieved greater success than states that have remained economically backward and traditional in culture.

## Conclusion

Within the last 50 years, two very different conceptions of demographic growth and its relationship to development have prevailed, leading to divergent positions on population policy. Within both points of view, however, demographic and policy objectives have been rooted in ideology and, in accordance with the idiosyncratic practices of the Mexican political system, have involved the participation of successive governments of the republic. The fundamental characteristic of the system is the centralization of power and the relatively limited participation of private citizens and local political groups. However, the participation of distinguished leaders from professional and academic fields has been important. Through their influence, pressure has been brought to bear upon the decisions of presidents and the groups of advisers around them.

The current Mexican population policy is confronted with several unresolved situations that have limited its capability to react rapidly and efficiently to the conditions and changes in national development strategy. The demographic changes that the population policy was designed to effect have been inadequate in relation to social and economic objectives established by successive development plans. The complex task of creating a mutually responsive relationship between demographic and economic trends is further complicated when overall sectoral or regional policies have been subject to frequent changes in their direction and reach. The integration of population variables into social and economic programs has not been achieved. Consequently, population programs have become independent, pursuing specific fertility and demographic growth objectives as goals, rather than as mechanisms for the attainment of social and economic change.

This situation can be observed in family planning programs, which, although relatively well integrated with health, education, and communication programs, are now reaching their limits in influence. Whether coverage can be raised to a higher level will depend on the creation of a comprehensive health policy, progress in social security, higher levels of female and male employment, and improvements in the overall social and economic factors that influence attitudes toward family formation and ideal family size.

The outcome of the current population policy with respect to fertility decline can be summarized as follows. The 1982 growth rate target of 2.5 percent was attained and surpassed. The 1. 9 percent growth rate target set for 1988 was not attained until 1990. This failure occurred because fertility did not decline as expected and coverage of family planning acceptors was below the intended level. The total fertility rate declined from 6.3 in 1973 to approximately 3.7 in 1987 (Dirección General de Planificación Familiar, 1989), but this level was considerably higher than the 3.2 children per family that had been targeted for 1987.

The reasons for the failure to attain this target are still uncertain. Possibly the economic crisis, which worsened after 1982, affected the organization and financing of family planning programs. Moreover, the reduction in fertility during previous years had taken place primarily among the urbanized middle and higher social classes. Smaller fertility declines have occurred among marginal urban groups and the rural sector, both of which populations have access only to limited health protection and deficient family planning services. Similarly, the effects of the early average age at which childbearing begins must be taken into consideration.

Finally, both political leaders and development planners need to become more deeply aware of the problems and challenges posed by Mexico's demographic trends. They must learn to think of population as an endogenous and interactive variable within the process of development (Cabrera, 1989). During the twentieth century, Mexico's demographic policies have undergone—and will continue to undergo—fundamental transformations in the effort to adapt them to the needs of the country's various development projects. Demographic changes should be more carefully tailored to accord with economic and development plans.

## Notes

1 The Mexican Secretary of the Interior and Foreign Affairs observed in 1838: "The openness and liberality with which our laws allowed and had favored the colonization of uninhabited areas of the Republic, instead of rendering the advantageous results promised by policy estimations for the establishment of such means of increasing the population, strength and wealth of nations, has brought about, due to our imprudent trust, countless evils and most regrettable consequences" (Sierra Brabatta, 1988: 7).

2 Greer (1966) considers the possibility of an overestimation in census data of 1910, and an underestimation in those of 1921. The author states that "the impact of the Revolution was that of holding the population of Mexico at a virtual standstill—an actual loss of 105,000 or less than 1 percent—during a period when it could otherwise have been expected to increase around two million. Although not causing a substantial absolute decrease of 800,000 as told in the census figures, the effect of the violent decades of the Revolution did prevent any increase, and thus interrupted an otherwise steady pattern of growth for twentieth-century Mexico" (pp. 116–117).

3 Ignacio García Téllez, Secretary of the Interior in President Cárdenas's government, commented: "I must emphasize that the demographic ideal which responds to our typical Mexican case lies mainly in the fact that the multiplication of our people should derive from its natural growth, even though it might be a slow process, as long as it constitutes a decisive testimonial to the improvement of human conditions and social progress; although it is desirable to have a numerous population, it is preferable to count on a unified and laborious community, capable of consolidating the welfare of our race and the plenitude of our nation" (Sierra Brabatta, 1988: 28).

4 See, for example, in *Informes Presidenciales* of 1958, 1964, and 1970, the speeches of Adolfo Ruiz Cortines, Adolfo López Mateos, and Gustavo Díaz Ordaz.

5 Conformity with presidential law was not always complete, however; in 1916 Felipe Carillo Puerto, governor of the state of Yucatán, encouraged the dissemination of Margaret Sanger's booklet.

6 Article 4 of the Constitution was amended as follows: "Women and men are equal before the law. The law shall protect the organization and development of families. All persons have the right to decide freely, responsibly and with knowledge on the number and spacing of children" (*Constitucíon Política de los Estados Unidos Mexicanos*, 1974).

## References


Alba, Francisco, et al. 1973. "Evolución demográfica de Mexico y políticas de población," *Salud Pública de México* época XV, 25, 1: 67–78.

Alemán, Miguel. 1947. *Primer Informe Presidencial*. Mexico City: Presidencia de la República.

Asociación Mexicana Población (AMEP). 1979. *Carta Informativa* no. 1. Mexico City: AMEP.

Benítez Zenteno, Raúl. 1967. *Análisis demográfico de México*. Mexico City: Instituto de Investigaciones Sociales, Universidad Nacional Autónoma de México.

Cabrera, Gustavo. 1966. *Diálogos sobre Población*. Mexico City: Centro de Estudios Económicos y Demográficos (CEED), El Colegio de México.

———. 1989. "Política de población: Un reto de Estado mexicano," *Demos. Carta demográfica sobre México* 2: 27–28.

———. 1990. "Políticas de población y cambio demográfico en el siglo XX," in *La sociedad mexicana en el umbral del milenio*, ed. Centro de Estudios Sociológicos (CES). Mexico City: El Colegio de México, pp. 249–272.

Centro de Estudios Económicos y Demográficos (CEED). 1979. *Dinámica de la población de México*. Mexico City: El Colegio de México.

Consejo Nacional de Población (CONAPO). 1980. *Política Demográfica Nacional: objetivos y metas*. Mexico City: CONAPO.

*Constitución Política de los Estados Unidos Mexicanos*. 1974. Mexico City: Editorial Themis.

———. 1990. Mexico City: Editorial Themis.

De la Peña, Moisés T. 1950. "Problemas demográficos y agrarios," in *Problemas aqrícolas e industriales de México*, 3 and 4. Mexico City: Talleres Gráficos de la Nación.

De Miguel, Amando. 1983. *Ensavo sobre la población en México*. Madrid: Centro de Investigaciones Sociológicas.

Dirección General de Estadística. 1934. *V Censo General de Población y Vivienda 1930*. Mexico City: Secretaría de la Economía Nacional.

———. 1948. *VI Censo General de Población y Vivienda, 1940*. Mexico City: Secretaría de la Economía Nacional.

———. 1952. *VII Censo General de Población y Vivienda, 1950*. Mexico City: Secretaría de Economía.

———. 1962. *VIII Censo General de Población y Vivienda, 1960*. Mexico City: Secretaría de Industria y Comercio.

———. 1973. *IX Censo General de Población y Vivienda, 1970*. Mexico City: Secretaría de Industria y Comercio.

Dirección General de Planificación Familiar. 1989. *Encuesta Nacional sobre Fecundidad y Salud*. Mexico City: Secretaría de Salud.

Durán Ochoa, Julio. 1955. *Población*. Mexico City: Fondo de Cultura Económica.

Echeverría, Luis. 1970. *Discurso de toma de posesión como presidente*. Mexico City: Presidencia de la República.

———. 1971. *Primer Informe Presidencial*. Mexico City: Presidencia de la República.

Episcopado Mexicano. 1973. "Mensaje del Espiscopadoal pueblo de México sobre la paternidad responsable," *Señal* 966, 8 September: central color pages.

Estados Unidos Mexicanos. 1989. "Constitución Federal de los Estados Unidos Mexicanos 1824," in *Las Constituciones de México*, ed. Comité de Asuntos Editoriales. Mexico City: Honorable Congreso de la Unión.

González Navarro, Moisés. 1960. *La colonización de Mexico 1877–1910*. Mexico City: Talleres de Impresión de Estampillas y Valores.

———. 1974. *Población y sociedad en México: 1900-1970*. Mexico City: Facultad de Ciencias Políticas y Sociales, Universidad Nacional Autónoma de México.

Greer, Robert G. 1966. "The demographic impact of the Mexican Revolution 1910–1921," Master's degree thesis, University of Texas, Austin.

*Informes presidenciales*. 1958. Adolfo Ruiz Cortines. Mexico City: Presidencia de la República.

———. 1964. Adolfo Lopéz Mateos. Mexico City: Presidencia de la República.

———. 1970. Gustavo Díaz Ordaz. Mexico City: Presidencia de la República.

Leal, Luisa María. 1975. "El proceso histórico de la Ley General de Población de México," Mexico City: Consejo Nacional de Población, mimeo.

Lerner, Susana. 1967. "La investigación y planeación demográficas en México," *Demoqrafía y Economía* 1, no. 1: 9–17.

Loyo, Gilberto. 1935. *La política demográfica de México*. Mexico City: Instituto de Estudios Sociales, Políticos y Económicos del Partido Nacional Revolucionario.

———. 1960. *La población de México, estado actual y tendencias, 1960–1980*. Mexico City: Editorial Cultura.

Márquez, Viviane. 1984. "El proceso social en la formación de políticas: el caso de la planificación familiar en México," *Estudios Sociológicos* 5 and 6: 309–333.

Meyer, Lorenzo. 1974. "El Estado mexicano contemporáneo," *Historia Mexicana* 23, 4: 722–752.

Nacional Financiera. 1978. *La economía mexicana en cifras*. Mexico City: Nacional Financiera.

Nagel, John S. 1978. *Mexico's Population Policy Turnaround*. Washington, DC: Population Reference Bureau.

Navarrete, Ifigenia. 1967. *Sobrepoblación y desarrollo económico*. Mexico City: Universidad Nacional Autónoma de México.

Sandoval, Alfonso. 1988. "La población en México," in *Mexico, Setenta y cinco años de revolución (desarrollo social I)*. Mexico City: Fondo de Cultura Económica/Instituto Nacional de Estudios Históricos de la Revolución Mexicana, p. 53.

Sierra Brabatta, C. J. 1988. "Antecedentes y comentarios a la Ley General de Población," unpublished manuscript.

Solís, Leopoldo. 1975. "Primer plan Sexenal (1934–1940)," in *Planes de desarrollo económico y social en México*. Mexico City: Secretaría de Educación Pública, Colección Sepsetentas.

Tamayo, Jorge L. 1962. "Geografía física," in *Geografía General de México*, vol. 1. Mexico City: Instituto Mexicano de Investigaciones Económicas.

Turner, Frederick C. 1974. *Responsible Parenthood: The Politics of Mexico's New Population Policies*. Washington, DC: American Enterprise Institute for Public Policy Research.

Urquidi, Víctor L. 1967. "El crecimiento demográfico y el desarrollo económico latinoamericano," *Demografía y Economía* 1, 1: 1–8.

———. 1969. "El desarrollo económico y el crecimiento de la poblacion," *Demoqrafía y Economía* 3, 2: 94–103.

———. 1973. "Problems of family planning in Mexico's population and development," *Gaceta Médica Mexicana* 105, 5: 475–511.

Warwick, Donald P. 1982. *Bitter Pills: Population Policies and Their Implementation in Eight Developing Countries*. New York: Cambridge University Press.

Zorrilla, Luis G. 1977. *Historia de las relaciones México–Estados Unidos, 1800–1958*, vol. 1. Mexico City: Editorial Porrúa.

# Islamic Doctrine and the Politics of Induced Fertility Change: An African Perspective

ALI A. MAZRUI

CONTRARY TO POPULAR BELIEF, Islam in Africa as a whole has been more relevant for the culture of lineage and procreation than for the culture of combat, more important in buttressing high fertility and defining lineage than as a *jihad*, or holy war, to eliminate rivals. This article concentrates primarily on Islam in sub-Saharan Africa (or black Africa).

The indigenous African religions are basically communal. Because they are not proselytizing religions, indigenous African creeds have not fought with each other. Over the centuries, Africans have waged many kinds of wars against each other, but rarely religious wars before the coming of Christianity and Islam (Mazrui, 1988; Mbiti, 1991).

Precisely because these two latter faiths were universalist in aspiration (that is, their followers sought to convert the whole of humankind), they were inherently competitive. The two Semitic religions are often rivals in Africa (Lapidus, 1990).

But why has Islam in Africa been less a cause of war than of impregnation, less a destroyer of rivals than a creator of new generations of believers? The main reason is that in its encouragement of large families (Mohsen, 1984; Ruthven, 1984; Patai, 1983; Weeks, 1988), the faith has reinforced indigenous African values. Yet, in its competitive tendency and universalist rivalry, Islam has run counter to the natural orientation of traditional African creeds, which are ecumenical and nonmonopolistic.

## Between the pacifier and the battlefield

In their relative support of fertility among believers, Islam and the indigenous African religions are basically allies. In certain versions of the jihad tradition, however, Islam diverges from the synthesizing tendencies of traditional African religions. Islam in Africa has been less a trumpet call to fanatical combat than an invitation to familial cohabitation.

But is the culture of procreation decisively distinct from the culture of war? Within the indigenous African paradigm, virility represents valor, and the sexual privileges of the warrior are inseparable from the culture of war among many groups, including the Nandi of Kenya and the Nuer of the southern Sudan (Huntingford, 1953; Hollis, 1909; Evans-Pritchard, 1933–35; 1940a and 1940b; Beidelman, 1958).

Part of the complexity lies in the historical division of labor between men and women, not just in Africa but almost everywhere. While men and women have had an almost equal share in determining births, men have had an overwhelmingly larger share in determining deaths. In twentieth-century Africa there has been no female equivalent of Idi Amin or Jean Bedel Bokassa or the original Boer architects of apartheid. Elsewhere in the world, there has been no female equivalent of Hitler, or Stalin, or Pol Pot.

Islam outside Africa has reinforced both the female role as childbearer and the male role as warrior (sometimes as the suicidal *mujahhid* destroying himself in combat). But within Africa Islam has more insistently reinforced the former than the latter.

Islam's culture of procreation has also included early motherhood. The proportion of women between 15 and 19 years old who are married in Muslim countries is 35 percent. In other developing nations the proportion is 19 percent. In developed countries the proportion is 3 percent (Weeks, 1988; United Nations, 1991).

A related cultural factor is the lineage system brought to Africa by the Arabs, which has accommodated considerable intermarriage. The Arab presence in Africa is concentrated in the northern part of the continent, along the Nile Valley, and, to some extent, along the East African coast. In these areas, the Arabo-Islamic lineage system has operated effectively for a long time: If the father is Arab, the child is Arab, regardless of the race or ethnic origins of the mother. Some ethnic groups (for example, Jews) throughout their history have theoretically protected themselves by strict endogamy and matrimonial exclusivity. Islam from the outset used exogamy as a strategy for making new ethnic alliances. The Prophet Muhammad set this precedent of diplomatic exogamy (Ruthven, 1984). Under certain circumstances, polygyny in Islam has also been among the factors that have encouraged large families.

In black Africa, more than anywhere else in the world, Islam and Christianity are rival religions, competing as alternative spiritual paradigms for the soul of a continent. In Lebanon, by contrast, Islam and Christianity are not attempting to convert each other, but are often, literally, at daggers drawn. In sub-Saharan Africa, followers of the two faiths often attempt spiritual conversion or reconversion, often scrambling for the allegiance of followers of indigenous African religions. Africa is still a free market of religions rather than a battleground for them (Sanneh, 1983).

However, if Islam and Christianity in Africa are in competition rather than at war, why did the first Sudanese civil war last from 1955 to 1972? Why was the second civil war partly precipitated by President Jaffer Numeiry's inauguration of Sharia'h (Islamic) Law in 1983? Why has the Muslim North of the Sudan been at war with the primarily non-Muslim South since the 1980s? Why was a coherent population policy impossible in the Sudan even in the years of relative peace between North and South from 1972 to 1983?

The divide between North and South in the Sudan is not just the difference made by Islam. The differences between the two regions go beyond religion, to the reality of ethnicity. The ethnocultural divide is both more comprehensive and deeper than the religious divide (Wai, 1981; Trimingham, 1980). Events in the Sudan since the resumption of the civil war there in the early 1980s have demonstrated that the country's problems are more complex than many assume (El-Affendi, 1990). Colonel John Garang's dissident Sudanese People's Liberation Movement (SPLM) and the Sudanese People's Liberation Army (SPLA) of the 1980s and 1990s were not secessionist movements like the Anya Nya of the earlier years. Nor were their supporters exclusively southern Sudanese (Khalid, 1987). In the North, Muslim "fundamentalists" have had determined (fellow) Muslim opponents. Likewise, the assorted tribes of the southern Sudan have, on the whole, manifested some unease about the demographic preponderance of the Dinka, some of whom are inhabitants of the northern Sudan (Mazrui, 1992). Attitudes toward family planning in the country generally are conditioned more by competitive ethnicity than by competitive sectarianism.

On the whole, religion in sub-Saharan Africa is probably divisive mainly when it reinforces a preexisting ethnic differentiation. At least, this is a hypothesis worth exploring. How was it possible for Senegal, which is more than 80 percent Muslim, to have had a Roman Catholic president, Léopold Sédar Senghor, for two decades without religious riots or the "need" for strict repression? In fact, Senegal since independence in 1960 has been one of the most open societies in sub-Saharan Africa. Senghor came from the Serer ethnic group, which comprises approximately 16 percent of the national population of Senegal. The majority of the Serer at the time Senghor served as president (1960–80) were Muslim, although a sizable Serer minority were Roman Catholic (Gellar, 1982). Why did not the 80 percent multiethnic Muslim population in Senegal revolt in indignation? In 1980, Senghor became the first African president to step down voluntarily as Head of State. He has been succeeded by a Muslim as Head of State, but President Abdou Diouf is married to a Roman Catholic; the First Lady of this Muslim society owes spiritual allegiance to the Vatican. In Senegal, Islam does not reinforce a preexisting ethnic divide.

In Nigeria almost all Hausa are Muslim; almost all Igbo are Christian; the Yoruba are split between Muslim and Christian. The Nigerian situation is midway between that of Sudan and Senegal. In Nigeria religion reinforces the ethnic divide between the Hausa and the Igbo. The Yoruba are potential bridgebuilders across the religious divide.

Debate about the shared tolerance of polygyny between Islam and African ethnocultures has been extensive. How does polygyny relate to kinship culture and comparative family values? Many Nigerian Christians are polygynous because of native custom. The rivalries about the number of children a woman should have, and the number of wives a man should have, do not translate neatly into a confrontation between Christians and Muslims (Caldwell, 1977).

## Population policy and competitive religion

In January 1988, President Ibrahim Babangida of Nigeria used the occasion of a budget speech to declare his government's intention to encourage and promote family planning and population control. In doing so, Babangida ignited a national debate. Particularly controversial was the government's tentative target of a maximum of four children per woman. The controversy concerned not merely the rights of women, but also the costs and benefits to different ethnic groups and rival religious denominations.

One major interpretation of the target of four children per woman was that it would favor the Muslims of Nigeria in their demographic rivalry with Christians. After all, Islam permits a maximum of four wives to every man. If each of those wives had the option of having four children, a Muslim man could theoretically father 16 children to every four children of a Christian marriage. Since President Babangida was himself a Muslim, there were suspicions in some Christian circles that his program of population control was intended to tilt Nigeria's demographic balance even more firmly in favor of Islam and its followers (Faruk, 1989). The last published and credible census in Nigeria, in 1963, had computed the Muslim proportion of the population as 47 percent, while Christians accounted for 35 percent. Since then, the Muslim percentage is widely believed to have exceeded the 50-percent mark (Faruk, 1989). By the mid-1980s, there were more Muslims in Nigeria than in any Arab country, including Egypt (Parrinder, 1992). Nigeria was not only the largest country in Africa, it also had the largest Muslim population on the continent, and may also have the largest Christian population as well.

Was the new population policy of the Babangida administration intended to consolidate and expand Islam's numerical advantage in Nigeria? Some Christian Nigerians were already worried that Islam in their country was expanding not only by conversion but also by faster procreation

(Elaigwu, 1984; Faruk, 1989; Goliber, 1989; Bisson, 1983). In such a situation, only a national policy that combined limiting every woman to four children with the principle of limiting every man to only one woman would avert an unfair war of procreation between Muslims and Christians. According to this school of thought, the struggle for population control in Nigeria would only work if it were combined with a crusade against polygyny.

When certain precepts of Islam are challenged, a Muslim defensiveness is provoked. Muslim fundamentalism relates itself to the culture of war. But when we look more closely, how does that orthodoxy relate to the culture of procreation?

Although Islam does not directly or totally prohibit contraception, the resurgence of Islamic fundamentalism in the Sudan, North Africa, and parts of West Africa has already had varied implications for fertility. Religious fundamentalism is, almost by definition, cultural revivalism. The role of women is redefined in ancient terms by the fundamentalist community. In Muslim Africa this has usually meant less choice for women in deciding how many children to bear, even when the family is middle class and relatively well educated. This diminishing of women's role is the result of Islam's deep cultural conservatism, its very old presence in Africa (much older than Christianity, except in Coptic Egypt and in Ethiopia), and its cultural comprehensiveness, entailing, as it does, more than religious or spiritual matters (Caldwell and Caldwell, 1985; Arens, 1973).

When Islamic fundamentalism is combined with the zeal and dedication of jihad (the mission of militant struggle in the path of Allah), that sense of mission often intensifies a cultural preference for male children as potential *mujahiddin* (crusaders for Allah). At that point, the culture of procreation merges with the culture of combat.

Muslim families in modern Africa are unlikely to resort to female infanticide simply because male babies are culturally preferable. But fundamentalist Muslim families in Africa and elsewhere may resort to having more and more children as a strategy of "multiplying in the name of Allah." Procreation can itself be counted as a form of jihad. The Muslim *ummah* (global community of Muslims) is allowed to expand by divine intervention and biological impregnation (Youssef, 1978).

But a third factor (after conservatism and the jihad complex) that may tend to make Islamic fundamentalism "pro-fertility" is a distrust of Western culture and the forces of Westernization in the Muslim world. Modern forms of contraception like condoms and intrauterine devices are identified as part of the Western culture of family planning. The financial and moral support that family planning clinics in Africa receive from Western sources deepens fundamentalists' distrust of Western-style contraception. The international family planning campaign, especially when backed by such citadels of Western power as the World Bank, may be perceived as a subtle invasion of the sanctity of the harem. In the eyes of the militant, contraception becomes a meeting point between the forces of cultural imperialism and the forces of preemptive genocide.

The fourth factor affecting the fundamentalist attitude to family planning is the association in Africa of the condom with prostitution. Only about one percent of the worldwide use of condoms occurs in Africa and the Middle East combined. The rubber device is widely assumed to be intended less for family planning than for protection against sexually transmitted diseases. Many African men, Muslim and non-Muslim, find the idea of using condoms with their wives distasteful. Indeed, many African women would be insulted if their men did so (Caldwell and Caldwell, 1985; Swartz, 1969). This association of condoms with prostitution and illicit sex may be particularly marked among African religious fundamentalists, both Christian and Muslim.

The fifth factor that makes Islamic fundamentalists distrustful of contraception is the presumption that birth control is a challenge to Allah's ability to feed all His children. The most distrusted rationale for contraception to a Muslim is often the economic rationale. This is because of a verse in the Qur'an denouncing economically motivated infanticide:

> And do not kill your children because of want or poverty. It is We who provide for you, and for them as well. Do not take away life which God has made sacred except in the cause of justice. Verily this is the injunction of God to you in the hope you will understand. ("Suratul-An'am" [Cattle], *The Qur'an*, v. 15)

Orthodox interpretations have concluded that the practice of family planning is morally dubious if the motive is to avoid poverty or avert a decline in the family's standard of living. On the other hand, contraception may be permissible when the reasons for it are connected with the health of the mother, or the risk of transmitting a major disease to the child, or when, because of the narrowness of the mother's pelvic girdle, damage could occur to the mother or the baby (Al-Kadhi, 1985).

However, contraception, even for economic reasons, is morally acceptable in Islam if the method used is withdrawal. In a *hadith* or tradition of the Prophet Muhammad, he is quoted as having said:

> There is no harm for a man to discharge semen outside the body of his wife if he desires no child. But Muslims should bear in mind that this notwithstanding, Allah creates whomsoever He intends to create. (Faruk, 1989: 10–11)

The fundamentalists among Muslims distrust many of these concessions to the use of contraception, especially in a world dominated by non-

Muslims and seemingly threatened by the forces of "evil" and "unbelief." The idea of small families among Muslim fundamentalists is, therefore, sometimes discouraged more for strategic rather than theological reasons. However, the theological concern remains that to reject the chance of an additional child is a form of ingratitude to Allah (Rosenthal, 1965; Rahman, 1980). The Qur'an itself proclaims that honest wealth and children are the supreme bounties of earthly life. To a fundamentalist, the rejection of a gift from God is a form of blasphemy.

## The anthropology of high fertility

Studies indicate that Muslims have the highest rates of population growth in sub-Saharan Africa. Factors that have favored higher fertility among Muslims include, first, an exceptionally early marriage for women, even by African standards. As has been noted, in Muslim countries 34 percent of brides are under the age of 20. Among the Hausa of northern Nigeria, brides below the age of 15 are much more common than they are among the Christianized Igbo in the south.[1]

Islamic African fertility rates have sometimes been seen as part of a global Islamic phenomenon, expansion not only by conversion, but even more by procreation:

> The single most remarkable demographic aspect of Islamic societies is the nearly universal high level of fertility—the average level of child-bearing in Islamic nations is 6 children per woman. . . . Fertility rates are highest for those Islamic nations in sub-Saharan Africa—an average of 6.6 births per woman. . . . Furthermore, African Islamic nations south of the Sahara have higher fertility on average than do the other developing nations in that region. (Weeks, 1988: 15)

Ironically, the lowest fertility rates on the continent today may well be in the almost purely Muslim north. This is because this most deeply Islamized part of Africa is also more susceptible than other parts of the continent to cultural Westernization. Geographically it is the closest part of Africa to Europe. Historically, the two regions have had old and deep ties. North Africa has also outstripped sub-Saharan Africa in levels of urbanization, industrialization, secularization, and Western-type education (Patai, 1993; Tibi, 1991; Ruthven, 1984). The Islamic hold on North Africa in such matters as procreation and family planning has probably been diluted by these processes in a manner akin to the weakening in the Western world of the influence of fundamentalist Christianity in this realm.

The second reason for higher fertility rates among Muslims in sub-Saharan Africa relates to levels of education. Although progress in women's education has been impressive in Africa in the second half of the twentieth century, there is still a smaller proportion of women who are highly educated among Muslims than among non-Muslims. Muslim families south of the Sahara are still profoundly distrustful of Western education as a potentially Christianizing, even corrupting influence on their womenfolk. Muslim girls are less likely to complete secondary education than are their Christian counterparts. In the absence of compulsory education, many Muslim families withdraw their girls as soon as they complete elementary school. Loss of access to education means Muslim women are less likely to be exposed to methods of contraception and the idea of birthspacing.

A third reason for the higher fertility rates among Muslims is the interplay between the Islamic patrilineal tradition and polygyny. Even among African groups that were previously matrilineal, Islam has strengthened descent through the father. A man with two, three, or four wives stands a chance of expanding the size of his clan. In conditions of politicized ethnic rivalry, this becomes a case of competitive procreation.

Arabized Sudanese from the north of the country have been known to vow not only to create more Muslims by conversion in the south but also to create more Arabs by intermarriage with black southerners. As noted earlier, by the lineage system of the Arabized north, the children would be Arab if the father was Arab—regardless of the mother's religion or ethnicity (Trimingham, 1980; Patai, 1983). "I am going south—to bring into the world 40 more Arabs before I die!" Such a boast is probably part of Sudanese competitive ethnicity. What has made such aspirations possible is the convergence of three forces: the strong Islamic patrilineal tradition, legitimized polygyny, and politicized ethnic rivalry.[2]

A fourth factor behind higher rates of fertility among Muslims in Africa may be Islam's undermining of traditional birthspacing practices. Under Islamic doctrine, breastfeeding is not the subject of sexual taboo. Moreover, Islam permits the baby of one mother to be fed regularly by another woman. The tradition of surrogate breastfeeding goes back to the experience of the Prophet Muhammad. Though this tradition was not unknown in black Africa, it acquired the status of a sacred precedent (*sunnah*) in parts of Islamized Eastern Africa.

This configuration of factors in Muslim Africa weakened the indigenous systems of birthspacing, which had relied on sexual abstinence for the mother during the period of breastfeeding (Doi, 1983; Davies, 1988).

A fifth factor is more elusive. Are Muslims more prone to fatalism? Do they more readily accept birth after birth than do Christians? What Laila S. El-Hamamsy said of Egyptian peasants may be true as well of Muslims south of the Sahara:

> It is not surprising that the Moslem peasant, living so precariously, should decide that fate, as determined by Allah, is responsible not only for the plea-

> sures and good things of life, but also for its misfortunes and pain. . . . Surely life would be insupportable were the peasant to have himself responsible by his own deeds and efforts for all that befalls him. (El-Hamamsy, 1972: 346)

The Kenya Fertility Survey in 1977 revealed that among Muslim married women in Kenya at the time, the average desired family size was 8.4 children. This was higher than for any other religious group. Among Catholic women the average preferred size was 7.1 children; among Protestants it was 7.0 children (Weeks, 1988).

John R. Weeks compares countries as well as religious groups on this issue. Weeks refers us to 1975:

> The Sri Lankan Fertility Survey had revealed that currently married Moslem women in that country expressed a desire for a family of 4.2 children. Buddhists preferred 3.9; Hindus preferred 3.8; and Christians expressed a desire for 3.4 children. Thus, although Moslem women preferred larger families than did members of other religious groups, it was also true that, regardless of religious preference, women in Sri Lanka wanted fewer children than did women in Kenya. (Weeks, 1988: 20)

In the Muslim world, the proportion of women using some method of contraception is 22 percent; in the rest of the developing world, it is 39 percent; and in the Western world, it is 70 percent (Weeks, 1988).

A sixth reason for high fertility among Muslims is the wide age difference between husbands and wives. A difference of eight years between a young bride and her older groom is fairly common in Muslim Africa. If the husband is considerably older than the wife, male dominance tends to be aggravated, and male dominance is associated with avoidance of contraception. A 1973 study in Senegal illustrated that the opposition of the husband was the second most important reason why women in Senegal did not practice family planning. The most important reason was "lack of information" (Nichols et al., 1985).

Sudanese society also illustrates that male dominance militates against contraception (Khalifa, 1988). Again the custom of a much older husband has been sanctified by the Islamic sunnah. After all, the Prophet's favorite wife, Aisha, was young enough to have been his daughter, as is common knowledge among modern-day Muslims.

But how does this explanation for high fertility relate to the historic divide between Sunnis and Shi'a in Islam? How does that divide affect the nature of Islamic fundamentalism and its orientation to birth control? Only a fraction of Africa's Muslim population is Shiite, the great majority being Sunni of different denominations (Shafii, Maliki, Hanbali, and Hanafi). However, since the Iranian revolution of 1979–80, proselytism in Africa in favor of Shiite conversions has expanded significantly. The Islamic Republic of Iran probably regards Africa as second only to the Middle East as an area "ready for Islamic resurgence." Subsidies from Iran for Muslim publications and activism in Africa increased dramatically in the 1980s. The Shiite direction from Teheran has recently become more visible internationally.

The minority status of Shiites almost everywhere in the Muslim world outside Iran, combined with their global status as a "faction" of the Islamic ummah, has helped to make the Shiites more eager for larger families than Sunni Muslims. Contraception in Shiite Islam is less directly forbidden than in the Roman Catholic Church, but it is also less actively practiced than among Sunni Muslims. In any case, there are only a very small number of Shiite Muslims in Africa (Trimingham, 1980; Parrinder, 1991).

## On prophecy and parenthood

Curiously enough, the most Westernized of all Muslim denominations also happens to be Shiite—the Ismaili followers of the Aga Khan (*Imam*) who live in East Africa as well as in South Asia. The wives (*begums*) of recent generations of Aga Khans have tended to be European women. The ruling families of this particular Shiite (Ismaili) denomination have practiced Western-style family planning. The general culture of the Ismaili denomination has been too Anglicized to encourage large families.

However, the followers of the Aga Khan in East Africa and elsewhere are more the exception than the rule in their attitudes toward fertility. In the African context, the Ismaili practice restrictive family planning more consistently than any other Muslim group, and more than Christian Protestant Africans as well.

In Kenya and Tanzania, Sunni Muslims have sometimes debated the issue of whether artificial contraception was consistent with the Sharia'h. There has been a consensus that withdrawal is legitimate. Debate has centered on the use of condoms instead. The Islamic chief judges of Zanzibar and Kenya have often reaffirmed the legitimacy of condoms. Since the end of World War II, almost all of half a dozen such chief *kadhis* in Kenya alone have given a *fatwa* (judgment) legitimizing family planning and confirming most forms of contraception. And every chief kadhi in Kenya and Tanzania has, of course, been a Sunni Muslim. However, the more fundamentalist Muslims in Kenya and Tanzania have often denounced condoms as instruments of *umalaya* (Swahili for prostitution).

Muslim children in Africa are taught quite early that among the glories of Islam in Arabia was the abolition of female infanticide. The Arabs before the advent of Islam had a concept of "surplus daughters." Pagan parents (especially the pagan father), having had "enough daughters already," would decide to use a pillow or a blanket to put an end to a new infant daughter (Rahman, 1986).

According to Islamic history taught in Qur'anic and Madrassa schools of Africa, it was the Prophet Muhammad who terminated this barbaric custom in Arabia. The question that has arisen since then in Africa is whether Islam's original crusade against Arab female infanticide should be extended doctrinally to include a ban on contraception in modern Africa. African Sunnism has agonized over the issue and decided that contraception is not a form of infanticide. Yet, the balance of sensibility in African Sunnism tends to be against abortion. Contraception is legitimated, but abortion is often regarded as much closer to the sin of infanticide.

As for the relationship between fertility and infant mortality, both Shiite and Sunni Islam are all too aware of how close the Prophet Muhammad came to having no descendants at all. Although the Prophet had several children (both male and female), only Fatma survived long enough to become a parent, the mother of Hassan and Hussein. Without Fatma there would have been no fusion between the blood of Muhammad and the blood of Ali. There would have been no assassination and martyrdom of Hussein at Karbala in AD 680, with all its implications for the martyrdom complex in Lebanon in the second half of the twentieth century. There would have been no Iranian Revolution, and no schism between Sunnism and Shiism in the grand history of Islam across the ages.

Because of the fate of the Prophet's own children, Islam is highly sensitized to infant mortality as a factor in its historical destiny. And yet historically in Shiite Islam, infant mortality deepened a distrust of contraception, while in Sunni Islam the mortality of the Prophet's children did not preclude the legitimacy of family planning in terms of doctrine.

The Ahmadiyya (Mirzai) movement in Africa (originally from South Asia) has often been the most militant of all the proselytizing Muslim efforts in Africa. The Ahmadiyya have been active in both West and Eastern Africa, converting more and more Africans to their school of Islam.

The Ahmadiyya movement, however small it may be as a faction in Muslim Africa, has succeeded in raising pressing issues of mortality, succession, and heredity. The relationship between demographic trends and social doctrine remains complex and sometimes inscrutable. Religion and culture have continued to condition family values, sometimes mediating between the competing claims of freedom of choice and social well-being.

## Conclusion

Islam originally arrived in Africa as a religion on the run. The first Muslim refugees arrived in Ethiopia early in the seventh century AD, fleeing from persecution in a polytheistic Arabia. The refugees presumably decided that intermarrying with Ethiopians and having large families was a way to strengthen their kinship ties in the Horn of Africa, and to expand their local Muslim communities in exile. The first great muezzin in the history of

Islam was Ethiopian—the famous Bilal, one of the Prophet Muhammad's favorite disciples. Islam's culture of procreation included racial intermarriage from the start.

The second significant arrival of Islam in Africa occurred in Egypt in AD 639. This time the Arabs arrived as conquerors rather than as refugees. But how was the Egyptian population not only converted to another religion (Islamization), but also transformed into Arabs (Arabization) over the centuries? How was this Arabization accomplished in the rest of North Africa and in much of the Nile Valley as well? Historically the culture of procreation among the conquering Arabs reinforced their jihad culture of combat. North Africa and much of the Nile Valley were Arabized as much by Arab intermarriage as by Arab conquest. The patrilineal system of the Arabs was part of their population policy, destined to transform a continent and affect the world.

By the twentieth century, Islam had changed Africa more through intermarriage and procreation than through war or combat culture. By the end of the twentieth century, half the population of Africa was professing Islam. The largest countries in Africa in population were, by then, Nigeria, Egypt, Ethiopia, and Zaire. Between them, these four countries accounted for more than 120 million Muslims.

But the lineage and procreation principles of Islam were not doctrinally isolated. They were profoundly affected by issues as varied as the shadow of infanticide and the economic justification for contraception. Preference for large families, or for male children, was often deeply rooted in the divergent origins of Sunni and Shiite Islam, and sometimes conditioned by the imperative of jihad. The culture of reproduction and the culture of competition have continued their dialectical interplay among the families of Islamized Africa.

## Notes

For consultation about Islamic doctrine, the author is particularly indebted to Sheikh Ahmed H. Al-Maamiry of Muscat, Oman, who in turn consulted the Grand Mufti of Oman and interpreted the *fatwa* (legal opinion) of such scholars as Sheikh Al-Shaarawy of Al-Azhar, Cairo, Egypt. The author is also indebted for theological assistance to his elder brother, Sheikh Harith Al'Amin Mazrui, in Kenya, and to Ahmed Abdullah, also of Kenya.

1 For this information, the author is indebted to research done for the BBC/PBS television series *The Africans: A Triple Heritage*.

2 See Mazrui, 1984, pp. 269–272; for other aspects of ethnic relations in Africa see, for example, in economic anthropology, Salem-Murdock (1989).

## References

Al-Kadhi, Ann Bragdon. 1985. "Women's education and its relation to fertility: Report from Baghdad," in *Women and the Family in the Middle East: New Voices of Change*, ed. Elizabeth Warnock Fernea. Austin: University of Texas Press, pp. 145–147.

Arens, W. 1973. "Mto Wa Mbu, a multi-ethnic community in northern Tanzania," in *Cultural Source Materials for Population Planning in East Africa* vol. 2: *Innovations and Communication*, ed. A. Molnos. Nairobi: East African Publishing House, pp. 442–447.

Beidelman, T. O. 1958. "Some Nuer notions of nakedness, nudity, and sexuality," *Africa* 38, 2: 113–132.

Bisson, Darlene Ferguson. 1983. *Demographic Factors and Population Activities in West Africa*. Abidjan, Côte d'Ivoire: United States Agency for International Development.

Caldwell, John C. 1977. "The economic rationality of high fertility: An investigation illustrated with Nigerian survey data," *Population Studies* 31, 1: 5–27.

———, and Pat Caldwell. 1985. "Cultural forces tending to sustain high fertility in tropical Africa," *PHN Technical Note* 85-16. Washington, DC: Health and Nutrition Department, World Bank.

———, and Pat Caldwell. 1987. "The cultural context of high fertility in sub-Saharan Africa," *Population and Development Review* 13, 3: 409–437.

Davies, Meryl Win. 1988. *Knowing One Another: Shaping an Islamic Anthropology*. London and New York: Monsell.

Doi, Abdur Rahman I. 1983. *Introduction to the Hadith*. Zaria, Nigeria: Hudahuda.

El-Affendi, Abdelwahab. 1990. "'Discovering the South': Sudanese dilemmas for Islam in Africa," *African Affairs* (London) 89, 356: 371–389.

El-Hamamsy, Laila S. 1972. "Belief systems and family planning in peasant societies," in *Are Our Descendants Doomed? Technological Change and Population Growth*, eds. Harrison Brown and Edward Hutchings Jr. New York: Viking Press, pp. 335–357.

Elaigwu, J. Isawa. 1984. "Nigeria's federal balance: Conflicts and compromises in the political system," *University of Jos Postgraduate Open Lecture Series* 1, 4: 1–50.

Evans-Pritchard, E. E. 1933–35. "The Nuer age sets," *Sudan Notes and Records, 1933–35*, vol. 19, part 2: 233–269.

———. 1940a. *The Nuer*. Oxford: Oxford University Press.

———. 1940b. "The political structure of the Nandi speaking people of Kenya," *Africa* 13, 3: 250–268.

Faruk, Alhaji Usman. 1989. *Family Planning: The Islamic Viewpoint*. Lagos and Kano, Nigeria: Paragon.

Gellar, Sheldon. 1982. *Senegal: An African Nation Between Islam and the West*. Boulder, CO: Westview Press.

Goliber, Thomas J. 1989. "Africa's expanding population: Old problems, new policies," *Population Bulletin* 44, 3: 1–50.

Grosse, Scott D. 1982. "The politics of family planning in the Maghrib," *Studies in Comparative International Development* 7, 1: 22–48.

Hollis, A. C. 1909. *The Nandi, Their Language and Folklore*. Oxford: Clarendon Press.

Huntingford, George W. 1953. *The Nandi of Kenya*. London: Routledge and Paul.

Khalid, Mansour (ed.). 1987. *John Garang Speaks*. London: KPI.

Khalifa, Mona A. 1988. "Attitudes of urban Sudanese men toward family planning," *Studies in Family Planning* 19, 4: 236–243.

Lapidus, Ira M. 1990. *A History of Islamic Societies*. Cambridge: Cambridge University Press, pp. 823–878; 890–898.

Mazrui, Ali A. 1984. *Africa's International Relations: The Diplomacy of Dependency and Change*. London: Heinemann Educational; Boulder, CO: Westview.

———. 1986. *The Africans: A Triple Heritage*. BBC/PBS television series. Executive producers: David Harrison and Charles Hobson.

———. 1988. "African Islam and competitive religion: Between revivalism and expansionism," *Third World Quarterly* 10, 2: 499–518.

———. 1992. "The Sudan in reverse: Northern separatists and southern integrationists," in *Culture and Contradiction: Dialectics of Wealth, Power and Symbol,* ed. Hermine G. De Soto. San Francisco: The Edwin Mellen Press, pp. 170–177.

Mbiti, John S. 1991. *Introduction to African Religion.* Portsmouth, NH and Nairobi: Heinemann.

Mohsen, Safia K. 1984. "Islam: The legal dimension," in *Islam: Legacy of the Past, Challenge of the Future,* eds. Don Peretz et al. New York: North River Press, pp. 99–128.

Nichols, Douglas et al. 1985. "Vanguard family planning acceptors in Senegal," *Studies in Family Planning* 16, 5: 271–278.

Parrinder, Geoffrey. 1992. "The religions of Africa," in *Africa South of the Sahara 1991.* London: Europa, pp. 90–94.

Patai, Raphael. 1983. *The Arab Mind.* New York: Scribner.

*Qur'an.* (*The Holy Qur'an,* translation and commentary by A. Yusuf Ali). Washington, DC: The Islamic Center.

Rahman, Afzular. 1986. *Role of Muslim Woman in Society*. London: Seerah Foundation.

Rahman, Fazlur. 1980. *Major Themes of the Qur'an.* Minneapolis and Chicago: Bibliotheca Islamica.

Rosenthal, E. I. J. 1965. *Islam in the Modern National State.* Cambridge: Cambridge University Press.

Ruthven, M. 1984. *Islam in the World.* New York: Oxford University Press.

Salem-Murdock, Muneera. 1989. *Knowing One Another: Shaping an Islamic Anthropology.* Salt Lake City: University of Utah Press.

Sanneh, Lamin. 1983. *West African Christianity: The Religious Impact.* London: C. Hurst.

Swartz, M. J. 1969. "Some cultural influences on family size in three East African societies," *Anthropological Quarterly* 42, 2: 73–88.

Tibi, Bassam. 1991. *Islam and the Cultural Accommodation of Social Change.* Boulder, CO: Westview Press.

Trimingham, J. S. 1980. *The Influence of Islam Upon Africa.* New York and London: Longman and Libraire du Liban.

United Nations. 1988, 1989, 1990, 1991, 1992. *Demographic Yearbook.* New York: United Nations.

Wai, Dunstan M. 1981. *The African–Arab Conflict in the Sudan.* New York: Africana.

Weeks, John R. 1988. "The demography of Islamic nations," *Population Bulletin* 43, 4: 1–53.

Youssef, Nadia H. 1978. "The status and fertility patterns of Muslim women," in *Women in the Muslim World,* eds. Lois Beck and Nikkie Keddie. Cambridge, MA and London: Harvard University Press, pp. 69–99.

# THE POLITICS OF POLICY FORMULATION AND IMPLEMENTATION: CASE STUDIES

# Two Kinds of Production: The Evolution of China's Family Planning Policy in the 1980s

TYRENE WHITE

THE ENACTMENT OF A SINGLE governmental policy rarely generates sustained international attention, but China's effort to enforce a one-child-per-couple family planning policy during the 1980s was a major exception. For two decades, China had propagated the use of birth control measures and encouraged fertility reduction. In the early 1970s, a two-child family became the model for nationwide emulation, and as the decade progressed substantial progress was made in reducing birth rates. By 1979, however, the post-Mao leadership was persuaded that China could not afford the luxury of even a two-child family; only a one-child limit, backed up by an aggressive campaign of propaganda, education, economic incentives, and administrative measures, could prevent China's population from burgeoning to catastrophic levels in the twenty-first century. Beginning in 1980, therefore, even rural couples were prohibited from having a second child, and by 1983 the state had tackled the *duotai* phenomenon (third or additional births) by promoting sterilization for childbearing-age couples with two or more children. After 1984, the intensity of the campaign was slowly scaled back; from 1985 to 1989, the birth limit for rural couples was progressively relaxed, and "single-daughter households" (*dunuhu*) were given permission to have a second child. By the end of the decade, China's policy remained the most aggressive and restrictive on record, but the policy had changed significantly since its inception.

In an effort to understand this policy and its implications, exhaustive research has been conducted on every aspect of the program. Despite the scope of that literature, however, no consensus has emerged on how to understand the process by which the policy was formulated and how it evolved. Where some see China's leaders making a rash decision to escalate pressures on couples to control births, others see a rational leadership responding to expert (albeit flawed) demographic advice and gearing population policy to fit modernization goals (Saith, 1984; Kallgren, 1985a). Where some see an autonomous state using threats and force to impose an unpopular and intrusive birth-limitation program, others see a constrained state, one in which the capacity to enforce strict birth limits is undercut by the impact of economic reforms (Aird, 1986; White, 1987). And where some see a core elite successfully wielding the tools of Maoism and Leninism to implement the one-child policy, others see a multi-tiered state organization in which local governments, exercising a degree of autonomy, have the capacity to modify state demands (Hardee-Cleaveland and Banister, 1988; Greenhalgh, 1990).

Despite their differences, however, these studies have three things in common. First, China's family planning and fertility-limitation programs are the central objects of analysis; the larger context of Chinese politics is considered only insofar as it is believed to impinge directly on the subject at hand. Second, assumptions about the nature of Chinese politics, how decisionmaking occurs, how the policy process works, and "how things get done" at the grassroots, are rarely subjected to serious scrutiny.[1] As a result, the place of the one-child policy in the broader public arena is sometimes exaggerated, and unexamined assumptions about the Chinese policy process significantly color the findings and analysis. Third, many assume that relaxations of the rural one-child limit are largely attributable to popular resistance.

The object of this chapter is to question each of these assumptions. To do so, the topic is approached from the opposite direction: (1) by placing the one-child policy on the periphery of Chinese politics, not at the center, and (2) by specifying at the outset four basic aspects of China's political and institutional climate that have provided the structure and context for decisionmaking on population policy in the post-Mao era. With that context in mind, the evolution of the one-child policy is reviewed, showing how changes in this policy sector have been the direct byproducts of broader developments in Chinese politics, not merely a response to popular resistance.

## The political economy of birth planning

During the period between 1977 and 1990, when the one-child policy was conceived and evolved, four sets of political and structural factors exercised a strong influence on the policy:

First, the policy evolved in a climate of comprehensive reform, a context that has specific ramifications for the policy process. In general, comprehensive reform initiatives are propelled by economic and political imperatives. Economically, they are usually designed to break out of a period of stagnation or crisis and to promote development. Politically, comprehensive programs help to distinguish new leaders from old ones and build support by promising across-the-board reforms that will solve old problems. New leaders therefore have a large political motive behind their initiatives, often leading to the construction of an ideology or political theory that demonstrates the coherence and interdependence of the reforms, specifies ambitious goals, and separates the regime irrevocably from its predecessor.

Inevitably, however, the package of initiatives must be broken down into individual policies with priorities attached. What results is a hierarchy of linked but unequal programs that vary over time with respect to the political attention and financial support they receive. That hierarchy is shaped first and foremost by considerations of power, that is, by what policies and programs are most crucial to the consolidation of the regime and the expansion of its political base. Since economic growth expands the resource pool available for distribution, economic programs usually become the privileged policy areas, while social policies that require short-term expenditures to achieve long-term, indirect economic gains occupy a lower rung on the policy hierarchy (Hirschman, 1973 and 1981; Grindle, 1980).

In the Chinese case, the post-Mao leadership defined economic modernization as the primary task of the party in the late twentieth century, staking its political legitimacy and prestige on its success in building the material base for socialism. In turn, the imperative of modernization led the state to tackle the problem of population growth with new vigor, calling on all cadres to "jointly grasp two kinds of production."[2] Behind the illusion of equal priorities imbedded in the slogan, however, is the shared understanding that population policy must conform to the needs of political and economic reform, not vice versa. Despite the urgency that was attached to the one-child policy, especially in the early 1980s, it was a *dependent* program—one that had to adjust to the needs of economic and political reform, not vice versa.

Second, political conflict, power shifts, and changes in party line may have significant effects on policies that stand far outside the arena of controversy. In this case, although some studies acknowledge the volatile political climate in which this policy emerged, the literature as a whole ignores the realities of elite conflict and political change in Beijing. Most analysts assume that the one-child policy emerged out of a leadership consensus (Banister, 1987); they therefore disregard the possibility that shifts in the composition of the central party and state organs could lead to modifications in the content or implementation of China's one-child policy. Nor is any connection made between subtle changes in the party line or political climate, and changes in the way population policy is publicly presented or actually implemented. Once again, this omission appears to be rooted in the assumption that state commitment to strict fertility control has insulated this policy from the real world of Chinese politics.

Third, like all public policy decisions, the decision to launch a one-child campaign was a decision to commit limited resources to a priority goal. And like all such decisions, it was subject to reconsideration and adjustment in the course of both routine and extraordinary economic and financial planning. In the Chinese context, the perennial problem of insufficient resources has been more or less constraining throughout the last decade, with budget deficits and financial shortfalls impeding many reform initiatives. Like other programs, the means through which the goals of fertility control could be achieved have been shaped by these financial constraints.

Lastly, behind the transient political and economic factors that may shape policy processes, enduring organizational features exercise a profound influence and resist change. In other words, structure counts. Policy initiatives take shape within state-specific structural and institutional configurations that turn general principles of bureaucratic politics into idiosyncratic patterns.

In China, the communist institution of centralized economic planning is the most important structural determinant of the policy process. As state planners work out the annual and five-year economic plans, their job is to find a fit between state goals and capabilities during a specified period of time, and to adjust investments and expenditures in ways that optimize economic performance while satisfying crucial political interests. Through this process, priorities are firmly established, regardless of external appearances and public statements. Moreover, extensive bargaining and haggling are involved among various bureaucratic actors, some defined by their functional mission (for example, the State Family Planning Commission or the Ministry of Health), and some defined territorially (for example, Hubei Province or Shanghai Municipality) (Lieberthal and Oksenberg, 1988). As China's reform process has introduced market mechanisms, the central planning process has been redirected toward central guidance, but the bureaucratic process whereby plans are devised remains essentially the same.

## The one-child policy revisited

How and in what way do these four aspects of policy context help to illuminate the politics of fertility policy? To answer that question, the post-Mao period may be divided into three phases: (1) the 1977–80 period of transition in the wake of Mao Zedong's death, when Hua Guofeng and Deng Xiaoping struggled to determine who would control China's destiny; (2) the 1980–84 period of rural reform, during which decollectivization and decommunization were undertaken; (3) the second stage of economic and political reform (1985–90), which can be divided into two parts—an early period during which the reformers sought to expand market mechanisms in the economy and reduce direct party interference in economic and administrative work; and a later period during which the reform coalition splintered under the weight of pressures for more far-reaching political reforms.

### Phase one: Leadership struggle and policy continuity (1977–80)

Between 1977 and 1980, crucial decisions were made to limit population growth and promote fertility control. Although these decisions have never been scrutinized with great rigor, two hypotheses that pertain to the decisionmaking process have gained currency in the literature. First, some have argued that the one-child policy was a radical escalation of China's birth-limitation program and goals, a reckless decision made by leaders intent on rapid modernization at all costs (Saith, 1984; Banister, 1987). Second, it is widely held that population policy stood above the fray of elite struggle; although Hua Guofeng, Mao Zedong's appointed successor, and Deng Xiaoping, Hua's practiced and skillful challenger, used other programs and policy areas to wage battles for political power, the need for population control was so compelling that a rare consensus emerged (Kallgren, 1985a; Banister, 1987).

To what extent do these hypotheses accurately reflect the decisionmaking process between 1977 and 1980? Despite the lack of access to senior decisionmakers who were involved in the process, the available data raise serious doubts about each of them.

*Radical escalation of goals* Although the one-child policy may well be considered a radical solution to China's population dilemma, the population goals and targets adopted between 1977 and 1980 were by no means radical policy departures. On the contrary, the goals were consistent with those articulated as early as the mid-1960s. What was new was the revelation that the goals could only be attained with a one-child limit, and the decision to push for universal and sustained compliance.

As early as 1965, population planning began to take root in China's overall economic planning process. In that year, Zhou Enlai articulated China's first long-range population goal in a speech on rural health work, calling for a reduction in the annual rate of population growth to "one percent before the end of the twentieth century," if not earlier (Zhou, 1985). Subsequently, Zhou included population targets in the draft of the Fourth Five-Year Plan (FYP), covering the period 1971–75, transferring population issues from the health arena to the planning arena (Liu Zheng, 1982). In practical terms this shift meant that the regime would apply to birth control goals the same target-oriented approach that had become a standard tool of policy implementation in the production process. Like material production, where numerical targets were routinely used to set output levels and standards of performance, quarterly, annual, and five-year targets for childbearing would be passed down to local cadres.

This ongoing policy planning was inherited by Hua Guofeng, Mao Zedong's appointed successor as party chairman. Under the Fourth FYP, the population growth rate targets for 1975 were set at "about 10 per thousand" in urban areas, and "under 15 per thousand" in rural areas (Fang et al., 1984). In 1975, Zhou's preliminary planning for the Fifth FYP (1976–80) set a national target of 14 per thousand for 1976 and 10 per thousand by 1980, precisely the goals articulated by Hua Guofeng in February 1978 (Ma et al., 1989). In addition, the initial draft of Hua's Ten-Year Plan for Economic Development called for a relatively modest goal of 9 per thousand by 1985, indicating a desire to stabilize population growth at under 10 per thousand.

In short, Hua Guofeng inherited a planning process already in motion, one that had been interrupted but not discarded during the turbulent years of 1976 and 1977. That planning process dictated to him a steady progression in the reduction of the population growth rate to less than one percent per annum, and it dictated the method by which this goal would be pursued at the local level. During this period, China's leaders were too preoccupied with who would win the battle for control of China's future to be intimately involved in a policy area so lacking in political significance. The built-in momentum of the central planning process drove decisionmaking on population goals during this period. As with other economic initiatives, a new and dramatic assault on the problem was deemed necessary to undo the harm wrought by the *Gang of Four* and its accomplices.[3] But the climate of emergency was a political one, as leaders and followers demonstrated through their optimistic assault on China's economic problems that they had drawn a political line between themselves and the Gang.

*The impact of elite conflict* Despite the shared concern of China's post-Mao leadership about the necessity of population control, it was political change that shaped decisionmaking during this period. From 1977 until mid-1979, when Hua Guofeng and his radical allies began to slip rapidly from power, Hua pursued a modified Maoist line in both political tone and policy content. A campaign to eliminate the "pernicious influences" of the Gang of Four was waged, a grandiose Ten-Year Plan for Economic Development was launched, and an atmosphere of mobilization permeated the party hierarchy. By December 1978, however, Deng succeeded in getting the first of a series of reform measures passed by the Party Central Committee, ushering in an era of reform that slowly deradicalized Chinese politics. For the time being, however, Hua and other conservatives continued to hold office and wield substantial power, creating a tense political atmosphere during 1979 and 1980.

The initial decisions to escalate China's birth-limitation program were made against this backdrop. In February 1978, Hua Guofeng called for reducing the annual rate of population growth to "under one percent within three years."[4] When the State Council Leading Group for Family Planning was revived in June, however, demographic projections showed that the

1980 target could be reached with a two-child policy, but the long-range goal of zero population growth could not. The Leading Group then recommended to the State Council a policy of promoting a one-child limit, allowing two at most, and strictly controlling third or additional births (Li, 1980).

Further decisions were not made until early 1979, after the Third Plenum of the Eleventh Central Committee in December 1978 tilted decisionmaking power in Deng's favor. The new economic line emphasized realistic planning over inflated and impossible goal-setting, and called for a three-year period of "readjustment and restructuring" to shift investment away from heavy industrial development toward other sectors of the economy. At the same time, Deng and his ally, veteran leader Chen Yun, repeatedly cited the poor living standards of China's huge rural population and the continuing problem of rural hunger, and called for improved standards of living (Ma et al., 1989). These themes persisted during 1979 and 1980, laying the foundation for the radical agricultural reforms to follow.

In the context of this economic adjustment and deficit-induced austerity, the projections on population growth began to assume new meaning. In January 1979, the State Council approved a "one is best, two at most" birth control policy, rewarding couples for limiting themselves to one child and penalizing them for third or additional births with fines and other economic sanctions (Liu Zheng, 1982). By June 1979, however, Hua Guofeng went further, lowering the population growth rate goal for 1985 from nine per thousand to the impossible level of five per thousand (Hua, 1979). While these targets could be justified in terms of China's long-term modernization goals, they were out of step with Deng's push for more realistic planning. As a result, these goals were quickly repudiated, along with Hua's grand development plan, and by September 1980 two new decisions had been made. On the one hand, the leadership announced the need for a "crash drive" to promote only one child per couple and hold population to 1.2 billion by the year 2000.[5] Provinces subsequently began to amend their regulations by imposing economic penalties on couples who gave birth to a second child, as well as a third child. On the other hand, the population target for 1981 was relaxed slightly, rising to ten per thousand from the original target of eight per thousand that had been set in November 1979. More significantly, the longer-range, more radical targets of five per thousand by 1985 and zero population growth by the year 2000 were quietly abandoned, and a policy loophole was created to enable couples with "practical difficulties" to bear a second child.[6]

This dualistic pattern—tightening limits on having a second child while adjusting and lowering longer-term goals—was driven by several considerations. First, the adjustment of the 1981 target was necessary to bring targets closer into line with likely outcomes; the failure to contain population growth within one percent by 1980 raised serious doubts about reducing it even further in subsequent years. At the same time, the reform coalition stood poised to promulgate a fundamental set of rural reforms that they hoped would do more than simply increase grain production. Their efforts were directed to improving rural living standards and rural incomes, a focus that would later be characterized by the slogan "it is glorious to get rich." For the time being, Deng gave this goal, like most others, a numerical target, calling for an increase in income per capita to US$1,000 by the year 2000. This ambitious target could be met only if an average of 1.5 children per couple were attained.[7] The rush was on, therefore, to set population targets at levels commensurate with Deng's economic vision for the turn of the century. Attention shifted toward the numerical total of 1.2 billion, which was used to set plans for provincial and lower-level administrative areas, and away from the questions of whether, how much, and how long population would continue to increase in the next century.

In short, despite the shared concern of China's post-Mao leadership about the necessity of population control, political change shaped decisionmaking during this period. When Hua Guofeng's radical approach to economic policy was undermined in 1979, his approach to population control was thrown out as well. Because China's birth planning limits tightened just as Hua Guofeng was losing ground, the subtle but important differences in approach to this issue have been overlooked. Faced with gloomy projections about the economic costs of childbearing, Hua's response was to use ultra-high planning targets (for example, holding the population growth rate to five per thousand by 1985) in order to pressure cadres for better results. Deng was equally intent on keeping birth rates low, but his overall policy line of improving rural livelihood, opposing mass mobilization, and placing economic pragmatism ahead of political orthodoxy left no room for Hua's brand of extremist planning. It did leave room for Deng's own brand, however—that of pursuing a one-child policy among China's 800 million peasants in order to achieve his ambitious economic goals by the end of the century.

### Phase two: Rural reform and party rectification (1980–84)

In the period 1980–84, rural China experienced a second transformation, as the new leadership systematically abolished the distinctive Maoist institutions and policies that had dominated rural life since the late 1950s. The transformation can be divided into three overlapping periods, each of which had corresponding consequences for the implementation of the one-child policy. The first period dates from September 1980 through 1983, when collective farming was slowly replaced by a contract system. Initially, this change met with much resistance and was implemented only in the poorest areas where the collective had been unable to provide for the population. From that foothold, it spread to middle- and upper-income areas over the next three years.[8]

The second period dates from 1982 through 1984, when communes were abolished and replaced with township governments. The short-term impact of this reform was more symbolic than real; nevertheless, many rural cadres were replaced by younger, more reform-minded leaders, signaling a generational change in grassroots politics.

The third period dates from 1983 through 1984, when the reformers pushed for rural diversification, the first step in the creation of China's "planned commodity economy." During this period, the reforms emphasized the development of rural sideline occupations and cash crops. In addition, peasants were allowed more freedom to determine what to plant and how to invest. The state also solidified peasant claims to the land they had contracted, guaranteeing 15-year leases to peasant contractors.[9]

Against this backdrop of radical reform, China's one-child policy took two crucial turns. In late 1982, a decision was made to promote sterilization for all couples under 40 years of age with two or more children.[10] As in previous mobilization campaigns, central goals were translated into local targets, that is, sterilization quotas were to be met by rural cadres. Second, a Central Document on birth planning was issued in early 1984; the document admitted that coercion was a serious problem in the implementation process, and repudiated the excesses that had characterized the sterilization campaign (Guojia jihua shengyu weiyuanhui, 1989).

Divergent interpretations of Central Document 7 may be found in the literature. One sees the document as a direct response to popular unrest over the sterilization campaign. Faced with a widespread peasant backlash, the state altered its tactics. Coercion was officially disavowed, and the percentage of couples who were allowed to have a second child increased slightly; nevertheless, the sterilization requirement continued to be implemented in a less provocative but extremely effective way. In this view, Central Document 7 did not signal an important turn in state policy. Instead, the document was simply good politics; at a time when the reform process was at a crucial juncture, it made sense to address peasant grievances at the highest level (Aird, 1986; Hardee-Cleaveland and Banister, 1988).

Another interpretation focuses on the context for implementing the one-child policy, arguing that structural and political changes made policy adaptation necessary. Structurally, decollectivization was complete by the end of 1983, and the reform of the commune system was nearly accomplished. The structural changes disrupted rural organization and distracted rural cadres, making it difficult to reward model one-child couples or penalize policy offenders. Moreover, measures designed to benefit one-child couples by allotting them more land, and to penalize additional births by withholding land or increasing mandatory grain delivery quotas, pitted the state's agricultural policy against birth control policy. Was a village cadre supposed to respect the 15-year land leases guaranteed by the state, or take land away from offenders and give it to one-child households?[11] Politically, the balance of power between cadres and peasants had also been altered by reform. Peasants were less vulnerable to local authority, better able to deflect or sidestep cadre demands, and more apt to vent their anger and hostility when provoked. This new political reality had fundamentally altered the context for implementation of the one-child policy, but campaign methods continued to be commonplace through 1983. As a result, cadre–peasant relations sunk to a new low, and cadres responsible for enforcement became fearful for their personal safety. In this context, Central Document 7 was issued; the document was the state's first serious effort to adjust its family planning program to a changing rural environment (Greenhalgh, 1990; White, 1990).

Both of these factors—peasant backlash and context for policy implementation—help to explain the new tone set by Central Document 7, but the broader context of Chinese politics and economics also played a crucial role. The politics of rural reform consumed the attention of the leadership during 1981–82. From the leaders' point of view, policy had been established in September 1980. Although the rural reforms were complicating enforcement, devising new implementation methods was the responsibility of the State Family Planning Commission (SFPC), not the Politburo. While the commission struggled to do its job, the policy line remained the same: Strict birth control was necessary to ensure the state's ability to achieve a rapid improvement in standards of living (Zhao, 1981).

The crucial question, therefore, was how to enforce the existing policy. By mid-1982, it was clear that the rural birth rates remained high and that one in every four births was third or higher parity (*duotai*). Since those figures seriously threatened the goals of the 1981–85 Five-Year Plan, the SFPC endorsed a new campaign targeted specifically at the *duotai* problem; all couples of childbearing age with two or more children were required to undergo sterilization.

The requirements of the central plan, plus the party line on agriculture, weighed heavily in making this decision. The final draft of the Sixth FYP was completed in November 1982, setting the goal for agricultural growth at 4 to 5 percent annually, and setting the population growth rate goal at an average of under 13 per thousand annually. By mid-1982, the prospects for meeting the agricultural target looked good, but population growth continued to top 14 per thousand and *duotai* continued to represent 24 percent of all births in 1982, down from 28 percent in 1981 but still too high (Liang and Peng, 1984). As a result, the leadership continued to fear that production increases, particularly grain production, would be undercut by population growth, slowing the rise in personal income that was necessary for economic expansion.

Although the sterilization campaign may have appeared to be a good means by which to achieve the goals of the plan—compensating for short-term organizational problems in the countryside and quickly lowering the *duotai* rate for one age cohort—in practice the campaign succumbed to a familiar pattern. Ambitious goals set by the center were pushed even higher by local officials seeking to demonstrate their enthusiasm, organizational ability, and political commitment. Under pressure to meet specific goals in a short period of time, all semblance of restraint was lost, and direct, physical coercion became an essential tool of policy implementation in many localities (White, forthcoming 1994).

The sterilization drive also proved to be expensive at a time when the reduction of "peasant burdens" (that is, taxes, levies, and fees of all sorts) was a major tenet of rural policy.[12] So serious was the concern over excessive financial burdens at the local level that in 1984 the key document on agricultural policy called for a reform of local finance. Township governments, successors to the now defunct communes, were to draw up fixed annual budgets under the guidance of higher-level governments. "All government-subsidized projects," including family planning, were to be "reviewed and reformed" in order to reduce expenses at the local level.[13] This reform, in short, would not permit government organs to impose on rural governments the kinds of excess expenditures incurred to implement the 1983 sterilization campaign. The explicit reference to family planning in this context was a clear signal that no matter how important birth control work continued to be, nothing was more important than the party's overall line on rural modernization.

Financial reforms were not the only important policy developments in late 1983 and early 1984. Agricultural policy focused on stabilizing the new system of household contracting by extending land contracts for 15 years, and on encouraging the development of rural industry and commerce. Rural entrepreneurs were permitted to hire rural labor, peasants were permitted to travel across provincial lines to sell their goods, and cadres were told to focus on serving the peasants, not ruling them. These policies served to boost agricultural production far beyond the expectations of the Sixth FYP, with agricultural output value increasing 11 percent in 1982, 9.6 percent in 1983, and an extraordinary 17.6 percent in 1984.

In addition, beginning in October 1983 the party undertook a rectification campaign designed to weed out all remaining supporters of Hua Guofeng's line, embrace younger, better-educated party members, and promote the development of a better party work style. A key focus of the campaign, which moved down the political hierarchy between 1983 and 1987, was on the rectification of party work style and methods of leadership; the goal was to eliminate "commandism" (arbitrary and dictatorial methods of rule) and to improve party–mass relations (Hu, 1982; Schram, 1984). With the economic reforms making old methods of leadership obsolete, unproductive, and dangerous to the party's grassroots support, improving cadre work methods became a prominent theme after 1983.

Within this broad context of a much-improved agricultural outlook and a push to improve the party's work style, Central Document 7 was transmitted in April 1984. The document, which was only available in excerpted and paraphrased form for many years, has been noted for its advocacy of "opening a small hole to close a big hole," or in other words, allowing more rural couples to have a second child in order to reduce the number of "illegitimate," unauthorized second or third births. The full text, however, is also notable for its tone of self-criticism centering on the problem of cadre work style in general, and coercion in particular (Guojia jihua shengyu weiyuanhui, 1989).[14] Noting that in conducting family planning work, "improving work style" and "improving party-mass relations" had been "neglected," the document went on to say:

> Concerning the phenomenon of coercion and violations of law and discipline that exist in some places, we have not adopted forceful remedial measures; in making plans for sterilization the demands were too urgent. . . . In those places where coercion exists, and no immediate solution has been found, the main responsibility is ours. We believed that because family planning work tasks are heavy, the appearance of coercion was unavoidable, and since only a minority use coercion, there was insufficient recognition of the danger and a lack of strong guidance. (Zhonggong guojia jihua shengyu weiyuanhui dangzu, 1984)

The document calls for a slight relaxation of the one-child limit in the countryside, increasing to 10 percent the proportion of rural couples to be permitted a second child, up from 5 percent in 1982; for further relaxations in the future as the rate of multiple births decreases; and for an end to the practice of "using one cut of the knife," that is, requiring uniformity in the sterilization of couples with two or more children.

In short, although policy changes at this juncture were no doubt motivated in part by a grassroots backlash over the sterilization campaign, as some have argued (Hardee-Cleaveland and Banister, 1988), popular displeasure with the campaign was not enough in itself to provoke a policy change. What mattered was the context in which that unrest occurred. Both the timing and the content of Central Document 7 demonstrate the heavy influence of the political line on party rectification, the unexpectedly rapid progress in developing the rural economy, and the shift in rural development strategy toward a service-oriented commodity economy. Because these three areas intersected on the question of how the party should treat the peasantry and how party cadres should conduct their work, family planning work was also affected. In future, cadres were to be judged by their work styles as well as their results.

### Phase three: The second stage of economic and political reform (1985–89)

During the second half of the 1980s the rural one-child limit was progressively relaxed, eventually allowing all "single-daughter households" (*dunuhu*) to have a second child after an interval of several years. Why this relaxation occurred has been a point of controversy. Some argue that peasant hostility and acts of retaliation left the state no choice but to loosen the policy. In this view, the relaxation was a tactical adjustment that did not diminish the extremity of China's birth-limitation program (Aird, 1986; Hardee-Cleaveland and Banister, 1988). Others argue that the process of reform made a strict one-child limit impractical and unenforceable; peasant households still relied on male labor power, and increased mobility among the rural population vastly complicated enforcement efforts. In addition, the decentralization of decisionmaking power that accompanied the process of reform increased the degree of slippage in implementation; provincial and local governments were in a better position to deflect central state demands, delaying the adoption of local regulations and crafting those regulations to fit local conditions (Greenhalgh, 1990; White, 1987 and 1991).

Both of these explanations for post-1984 policy evolution have merit. Once again, however, each concentrates on the immediate and proximate causes of policy change, neglecting the broader political and economic context in which these changes occurred. Three interrelated aspects of that context will be reviewed here: first, the effects of rapid economic growth in the rural sector; second, the effect of ongoing economic reforms on the planning process; and third, the country's changing political and economic climate.

By the end of 1984, the first phase of agricultural reforms had been completed. In 1985, therefore, the state took the crucial step of abolishing the quota system for most agricultural products, including grain, replacing it with a purchasing system based on guaranteed prices for contracted products and negotiated terms for the purchase of excess production (Oi, 1989). With this step, plus a simultaneous reform in the industrial sector, the state found itself halfway between plan and market, still deeply involved in the planning and management of the economy, but dependent on market forces as well.

Despite difficulties in introducing and implementing the reforms (including inflation, transport problems, and cadre corruption), by the end of 1985 the economic reforms were proclaimed a resounding success, with all of the goals of the Sixth FYP fulfilled or exceeded. The growth rates for agriculture and industry far outstripped the planned rates, and rural industry expanded at an extraordinary rate, absorbing 60 million laborers by 1986

(Xinhua Domestic Service, 1986). This development helped to lessen the problem of rural unemployment and underemployment, a key concern of the planners. At the same time, living standards increased significantly, particularly in the countryside, where inflation-adjusted incomes rose by 13.7 percent annually (Zhao, 1986).

This economic success played an important role in recasting the terms of debate on population control. Since 1979, family planning propaganda had stressed that population growth was a serious threat to China's economic development, not only in the short term (through the end of the century), but over the long term as well. To remove that threat, a strict one-child policy was necessary. By 1986, however, a more widely held and less controversial view began to gain legitimacy—namely, that economic development was the key to reducing birth rates, not vice versa.

The influence of this perspective can be seen in the Seventh FYP (1986–90), which was approved in April 1986, and in a new document on family planning that was issued in May. Under the Sixth FYP, both the plan and senior leaders made clear that population control was directly linked to the key goal of rapidly improved living standards and incomes, and discussions of population control were combined with discussions of improving income levels. In contrast, the Seventh FYP and Zhao's report deemphasize that relationship. Both discuss population control under the less distinguished category of "other social programs," with Zhao's report simply mentioning family planning in passing. The plan refers to family planning twice, once under "other social programs," where the goal of "trying" to keep the annual population growth rate at "around 12.4 per thousand" is specified, and again under a section on "People's Life and Social Security," in which the population goal for 1990 is set at 1.113 billion.[15] Despite the rigorous goal of 12.4 per thousand, the failure to emphasize family planning or urge better efforts stands in striking contrast to the tone and content of the 1980–85 period.

That change of tone also permeated Central Document 13 (1986) on family planning (Guojia jihua shengyu weiyuanhui, 1989).[16] While reiterating the importance of population control in promoting economic development, the standard argument, the document legitimized for the first time an alternative view:

> The condition of individual localities having differing birth levels illustrates that unequal economic and cultural development influences birth levels, reflecting the important function of economic development, culture and education in transforming birth rates. . . . (Guojia jihua shengyu weiyuanhui, 1989)

In other words, economic development was a cause, not merely a consequence, of declining birth rates. This perspective made the case for flexible and more relaxed rural implementation more compelling, paving the way for a series of policy concessions between 1986 and 1989.

If the changed economic climate played a key role in recasting the framework for China's birth-limitation effort after 1985 (at least in the countryside), the deteriorating political and economic climate after 1986 had a similar effect on the policy-implementation process. When the 13th Party Congress convened in October 1987, several developments recast the context within which family planning policy was debated. In the aftermath of student demonstrations in Beijing in December 1986, General Secretary Hu Yaobang was forced to step down from his post. He was succeeded by Zhao Ziyang in the fall, while Li Peng was promoted to succeed Zhao as premier. Also, although the economy continued to grow rapidly, the state had lost the ability to rein in spending and investment by local governments, which resulted in a pattern of overheated growth and rising inflation. With the state monopoly of supply and distribution abolished, corruption flourished, as cadres and entrepreneurs captured large supplies of key materials (for example, fertilizer and diesel oil) and sold them at inflated prices during peak demand. In addition, the abolition of state grain procurement had caused grain production to stagnate, since many peasants chose to shift acreage from grain to more lucrative crops.

These economic problems fueled a growing political debate between the "radical reformers" and the "central planners" among the Chinese elite. As early as 1983, the latter group had expressed its concern over "spiritual pollution" and attempted to emphasize Leninist discipline and orthodoxy as an antidote to "polluting" Western influences. Those concerns heated up substantially after 1984, and by late 1986 serious problems with crime, corruption, poor party discipline, and social unrest in the form of student demonstrations primed them for an attack on the reforms (Sullivan, 1988). After granting his support for an assault on "rightism" and "bourgeois liberalization" in early 1987, however, Deng Xiaoping reversed himself in April, a move that spared him any major retrenchment on economic reform, but left unresolved the question of political reform (Schram, 1988).

In that context, Zhao Ziyang delivered a speech on "socialism with Chinese characteristics" at the 13th Party Congress in October 1987. The speech reaffirmed the policy of reform as a necessary condition for China's modernization. Zhao also argued that China was in the primary stage of socialism, and that it would take 100 years, dating from the initial socialist transformation in the 1950s, to achieve full socialist modernization. With respect to family planning, the speech continued the trend set in motion with the Seventh FYP, grouping family planning with environmental protection as an important but secondary "social program." In addition, Zhao's speech was the first to discuss the problem of an aging population, suggesting a need to balance short-term pressures to reduce birth rates against the long-term consequences of a sudden demographic shift to one-child families (Zhao, 1987).

The general policy set in motion in 1984 was thus continued into 1988, but the debate over correct policy became more intense as birth rates began to rise. One article in a leading newspaper argued that the relaxation of rural limits on having a second child was a mistake, and that senior policymakers were directly to blame (Liu Jingzhi, 1988). Another newspaper ran an article far more blunt than the usual fare. The author argued that it was not enough to rely on "feeble and gloomy propaganda and education," to enforce birth limits; he called for the use of "coercion and control" before it was "too late" (Xie, 1989). At the same time, other specialists called for a two-child policy, while still others supported the current policy as the best compromise (Peng, 1989).

In the end, the established policy prevailed. In his government work report for 1988, Premier Li Peng placed more emphasis on the "strict" control of population growth than had Zhao Ziyang six months earlier. Speaking at a meeting for directors of provincial family planning commissions, however, the new head of the SFPC, Peng Peiyun, refuted the charge of poor policymaking, arguing that the inadequate performance in 1986 and 1987 was "a work problem, not a mistake in policy" (Peng, 1989).

What accounts for China's restraint when its leaders were faced with rising birth rates and growth rates that exceeded national plans and threatened the long-term goals? First and foremost, China's leaders remained concerned that the policy be tolerable to the peasantry, on whom they depended for economic development and for political support, and to rural cadres, whose cooperation was sorely needed in the developmental effort. By 1988, conditions in the countryside were already tense, and family planning policy was only one of many festering issues. Peasants complained about cadre corruption, "commandism," and nepotism, while cadres complained that they were caught between angry peasants and their superiors, unable to meet anyone's demands but blamed for all that went wrong (White, 1990). Increasingly, this frustration was spilling over into violence. In a single county in Jiangsu, 381 "incidents of revenge" were reported within a year and a half in just 12 towns and townships. Of the five categories into which these incidents were divided, the largest category was related to the enforcement of family planning, comprising 32 percent of all incidents. Even conflicts over grain requisitions fell second to family planning (Su, 1988). In that climate, a new campaign would only cause the situation to deteriorate, something that reformers intent on implementing market-oriented urban reforms could ill afford.

In purely economic terms, mobilization would have been a very expensive proposition in the late 1980s, and by 1988 money was in short supply. Funds for family planning began to fall off under the Seventh FYP. In 1988, Li Peng responded to demands for more funding by noting that the center was having financial difficulties of its own, and encouraging local governments to assume more of the burden (Guojia jihua shengyu weiyuanhui, 1989). Because of the reforms, however, local governments controlled their own purse strings; they were reluctant to invest more in a state-mandated family planning program that generated no revenue.

Most fundamentally, having gone far toward eroding the mechanisms of central planning that kept grassroots units under strict central supervision, the center could not readily impose a mobilization campaign to enforce family planning. Economically, the center no longer directly controlled the profits derived from local enterprises, and state policy encouraged investment of those revenues in profit-making ventures, not in costly family planning efforts. Politically, it had spent five years rectifying a party work style based on centralization, control, administrative orders, and coercion, in order to implement market reforms. Having shifted its organization toward a guidance approach to management and having abolished the practice of assigning detailed targets to local units, the center was in no position to reinstate them for a singleminded purpose. So doing would take another more fundamental change in party line.

That change came in June 1989, with the decision to have the army open fire on pro-democracy demonstrators in Beijing. In its immediate aftermath, the line of the central planners was promoted, and media reports advanced conservative views. Despite the threat to retreat from reform and restore centralized economic planning, however, conservative leaders were thwarted by opposition from local governments. By the end of 1990, when the draft of the Eighth FYP (1991–95) was completed, it was clear that the reform process would go cautiously forward.

Similarly, there was no major retreat from the previous policy on family planning, despite some reactionary rhetoric in late 1989. Politically, the center was in no better position to clamp down on birth control than it had been before the crisis. On the contrary, assuring political stability in the countryside and improving cadre–mass relations became central tenets of party policy in the fall of 1989 and spring of 1990 (Crook, 1990). Economically, neither the central planners nor reformers needed a sullen peasantry at a time of political and economic crisis. Rural stability had become a strategic issue, one that took precedence over concerns for population growth and militated against any immediate change in policy.

The results of the July 1990 census began to alter that calculus, however. The census revealed that the population targets for the Seventh Five-Year Plan had been exceeded by large margins. It also revealed the full extent of underreporting in the statistical system. As it entered the Eighth Five-Year Plan period (1991–95), therefore, China had a population count of 1.143 billion. But the worst was yet to come. In the first half of the 1990s, the average number of women entering their childbearing years (ages 15–49) would increase by 8 percent per annum compared with the previous plan period, averaging 322 million per year. Similarly, the number of women in their peak childbearing years (20–29) would increase by 16 percent, and reach 122 million annually (Chen, 1991). The number of births was expected to top 23 million per year.

These disconcerting figures, combined with the failure to meet planning targets for the 1986–90 period, had a predictable effect. In late 1990 and 1991, as the Eighth Five-Year Plan began to take effect, the rhetoric about population control began to escalate. A new central document on family planning was published, and a new campaign was launched emphasizing the need for strict enforcement of the existing policy.[17] Cadres were placed on alert that they had to take family planning targets more seriously, and pressures increased to meet new, stricter birth quotas. In the short run, the new mobilization resulted in significantly lower birth rates and population growth rates during 1991 and 1992, as cadres worked to keep within the limits of their new planning guidelines.

## Conclusion

The slogan that Chinese officials have used to link family planning to economic production—"jointly grasp two kinds of production"—embodies the essence of China's approach to the question of birth control, and reflects the political and structural reality of a Leninist party and a communist state. In a system where the state is organized on the assumption that centralized economic and social planning under the leadership of the Communist Party is both necessary and correct, policymaking is by definition comprehensive. Key economic requirements and goals set the agenda for the rest of the society and economy, making social programs like family planning structurally and politically dependent on the state's prevailing economic line and initiatives. Under a Leninist, centrally planned system, strong party discipline and political vigilance ensure that local governments and grassroots cadres will do a reasonably good job in responding to the second-tier priorities, maintaining the organizational illusion that under socialism all goals are equally important and worthy of equal effort. When a Leninist state introduces market reforms, as in China, momentum must still be created at the center. The new party line therefore shifts to one of reform, new and interrelated priorities are established, and the state sets about a new pattern of social engineering under the "unified party line."

Once market reforms are introduced, however, the illusion of sustained and equal attention to all programs on the agenda is much harder to maintain. In China, the effect of successful reform was to encourage localities to make their own decisions and pursue their own economic interests (within limits defined by the center); the reforms also gave them the resources to carry through local projects. Under those circumstances, the fiction of giving equal play to "two kinds of production" could not be maintained. Caught between its overall policy line on the peasantry and agricultural development and the need for strict control of population growth, the state moved toward a compromise position halfway between plan and market—one that preserved centralized population planning and local pressures for compliance, while conceding implicitly that the best long-term route to fertility reduction was through a market-driven process of rural development.

Even as China moved into the 1990s, however, the centralizing and planning impulses that are endemic to the country's political system continued to produce a cyclical pattern. At the beginning and end of each five-year plan, "population production" targets, like material production targets, are made the paramount indicators of local performance, and become the basis of cadre evaluation. Under such a system, that a new campaign will be launched to meet the necessary targets becomes predictable and routine, just as predictable and routine as it is that the failure to meet targets in any one cycle will cause efforts in the succeeding planning period to be redoubled. That this pattern continues to hold in the 1990s, however, is not evidence of a retreat from the compromise of the 1980s, but is evidence, rather, of an effort to define clearly and enforce rigorously the limits of that compromise. Even if China is able to meet its population goals for the 1990s, an average of 23 million births per year during that period is anticipated, and total population is expected to climb above 1.3 billion. Similarly, the country's 1990 workforce of 727 million will grow to 858 million by the year 2000, and continue to rise to 977 million by 2020 (Xinhua Domestic Service, 1991). Faced with these disturbing realities, the government has concluded that there is no alternative to tighter rural enforcement during the peak period of the 1990s.

## Notes

Research for this article was supported by the Swarthmore College Faculty Research Support Fund; a postdoctoral fellowship from the Fairbank Center for East Asian Research, Harvard University; and the Committee on Scholarly Communication with the People's Republic of China, US National Academy of Sciences.

1 An important exception is Greenhalgh, 1990.

2 The slogan, "jointly grasp two kinds of production," is rooted in the Marxist notion that the production of the means of subsistence and the reproduction of human beings are two halves of a single historical process.

3 The "Gang of Four"—Jiang Qing (Mao Zedong's wife), Zhang Chunqiao, Wang Hongwen, and Yao Wenyuan—were radical leaders during the Cultural Revolution. They were dubbed the Gang after Mao's death in 1976, when they were arrested and made scapegoats for the crimes and abuses of the decade of the Cultural Revolution.

4 The full text of the report may be found in *Beijing Review* (Hua, 1978).

5 See Hua Guofeng's speech to the third session of the fifth National People's Congress, *Beijing Review* (Hua, 1980).

6 In addition to Hua's speech, which does not refer to a goal of zero population growth, see Yao Yilin's report (1980) on the national economic plan for 1981, and the "Open Letter" on family planning to all party cadres and youth league members (*Renmin Ribao*, 1980).

7 See, for example, the discussion in Yu, 1980.

8 For a description of the politics of this process, see Zweig, 1987.

9 On the policy in 1984, see Kallgren, 1985b; Lieberthal, 1985; Stone, 1985; and Kueh, 1985, all in a single issue of *China Quarterly*.

10 For a discussion of how this decision was reached, see White, forthcoming, 1994.

11 For a complete description of these difficulties, see White, 1987.

12 In the city of Wuhan, for example, the annual budget for family planning in 1983 was 600,000 yuan. However, the sterilization campaign alone cost one million yuan, and was paid out of funds held by the municipal finance bureau. (Personal interview with municipal family planning officials in Wuhan, 1984.)

13 For the translated text of the document, see *China Quarterly*, 1985.

14 For a complete translation of this and other key documents, see White, 1992.

15 For the text of the Seventh Five-Year Plan, see Zhao, 1986.

16 The document's correspondence with the national plan is clear from its content, but also from its date: 23 March 1986. Zhao Ziyang delivered his National People's Congress report on the Seventh FYP on 25 March 1986.

17 Evidence of the new campaign could be seen in the summer of 1991. In June and July of that year, the author saw new banners and freshly painted family planning signs in rural towns and villages across Hebei, Hubei, and Shandong.

## References

Aird, John S. 1986. "Coercion in family planning: Causes, methods and consequences," in United States Congress Joint Economic Committee, *China's Economy Looks Toward the Year 2000*, Vol. I: *The Four Modernizations*. Washington, DC: United States Government Printing Office.

Banister, Judith. 1987. *China's Changing Population*. Stanford, CA: Stanford University Press.

Chen Jian. 1991. "Bring about coordination between population growth and economic development," *Liaowang* (*Outlook*) 52 (24 December 1990), in *Foreign Broadcast Information Service—Daily Report, China*, 31 January 1991, pp. 38–39.

*China Quarterly*. 1985. "*Documentation*: Circular of the Central Committee of the Chinese Communist Party on rural work during 1984," *China Quarterly* 101: 132–142.

Crook, Frederick. 1990. "Sources of rural instability," *China Business Review* 17, 4 (July–August): 12–15.

Fang Weizhong, Wang Renzhi, Du Shipu, and Liu Suinian (eds.). 1984. *Economic Chronicle of the People's Republic of China, 1949–1980* (in Chinese). Beijing: China Social Sciences Press.

Greenhalgh, Susan. 1990. "The evolution of the one-child policy in Shaanxi," *China Quarterly*, no. 122 (June): 191–229.

Grindle, Merilee S. (ed.). 1980. *Politics and Policy Implementation in the Third World*. Princeton, NJ: Princeton University Press.

Guojia jihua shengyu weiyuanhui. 1989. *Selected Important Documents on Family Planning Since the Third Plenum of the Eleventh Central Committee* (in Chinese). Beijing: Central Party School Press.

Hardee-Cleaveland, Karen, and Judith Banister. 1988. "Fertility policy and implementation in China, 1986–88," *Population and Development Review* 14, 2: 245–286.

Hirschman, Albert O. 1973. *Journeys Toward Progress: Studies of Economic Policy-Making in Latin America*. New York: W.W. Norton.

———. 1981. *Essays in Trespassing: Economics to Politics and Beyond*. Cambridge, England: Cambridge University Press.

Hu Yaobang. 1982. "Create a new situation in all fields of socialist modernization," *Beijing Review* 25, 37: 11–40.

Hua Guofeng. 1978. "Unite and strive to build a modern, powerful socialist country!," *Beijing Review* 21 (10 March): 7–45.

———. 1979. "Report on the work of the government," *Beijing Review* 22, no. 27: 5–31.

Kallgren, Joyce K. 1985a. "Politics, welfare, and change: The single-child family in China," in *The Political Economy of Reform in Post-Mao China*, eds. Elizabeth J. Perry and Christine Wong. Cambridge, MA: Council on East Asian Studies, Harvard University, pp. 131–156.

———. 1985b. "The concept of decentralization in Document No. 1, 1984," *China Quarterly* 101: 104–108.

Kueh, Y.Y. 1985. "The economics of the `Second Land Reform' in China," *China Quarterly* 101: 122–131.

Li Xiuzhen. 1980. "The current situation and tasks in family planning work" (in Chinese), *Renkou Yanjiu (Population Research)* 1: 3–5, 47.

Liang Jimin and Peng Zhiliang. 1984. "Understand and implement the party's family planning policies in an all-round and accurate way" (in Chinese), *Renkou Yanjiu (Population Research)* 3: 11–15.

Lieberthal, Kenneth. 1985. "The political implications of Document No. 1, 1984," *China Quarterly* 101: 109–113.

———, and Michel Oksenberg. 1988. *Policy-Making in China: Leaders, Structures, and Processes*. Princeton, NJ: Princeton University Press.

Liu Jingzhi. 1988. "Experts concerned are not optimistic about China's population situation, and think that interference by officials is an important reason why birth rate has risen again," *Guangming Ribao (Bright Daily)*, 6 March, p. 2, in *Foreign Broadcast Information Service—Daily Report, China*, 18 March, pp. 14–15.

Liu Zheng. 1982. *China's Population* (in Chinese). Beijing: Renmin chubanshe.

Ma Qibin et al. (eds.). 1989. *Forty Years of Rule by the Chinese Communist Party* (in Chinese). Beijing: Central Party History Materials Press.

Oi, Jean C. 1989. *State and Peasant in Contemporary China: The Political Economy of Village Government*. Berkeley: University of California Press.

Peng Peiyun. 1989. "Speech at the closing of national family planning commission directors meeting," in *Selected Important Documents on Family Planning Since the Third Plenum of the Eleventh Central Committee* (in Chinese). Beijing: Central Party School Publishers, pp. 108–120.

*Renmin Ribao (People's Daily)*. 1980. "Open letter." 22 September.

Saith, Ashwani. 1984. "China's new population policies," in *Institutional Reform and Economic Development in the Chinese Countryside*, ed. Keith Griffin. Armonk, NY: M. E. Sharpe, pp. 176–209.

Schram, Stuart R. 1984. "Economics in command? Ideology and politics since the third plenum, 1978–84," *China Quarterly* 99: 437–461.

———. 1988. "China after the 13th Congress," *China Quarterly* 114: 177–197.

Stone, Bruce. 1985. "The basis for Chinese agricultural growth in the 1980s and 1990s: A comment on Document No. 1, 1984," *China Quarterly* 101: 114–121.

Su Suining. 1988. "There are many causes of strained relations between cadres and masses in the rural areas," *Nongmin Ribao (Farmers' Daily)*, 26 September, p. 1., in *Foreign Broadcast Information Service—Daily Report, China*, 7 October, p. 13.

Sullivan, Lawrence R. 1988. "Assault on the reforms: Conservative criticism of political and economic liberalization in China, 1985–86," *China Quarterly* 114: 198–222.

White, Tyrene. 1987. "Implementing the `one-child-per-couple' birth control policy in rural china: National goals and local politics," in *Policy Implementation in Post-Mao China*, ed. David M. Lampton. Berkeley: University of California Press, pp. 284–317.

———. 1990. "Political reform and rural government," in *Chinese Society on the Eve of Tiananmen*, eds. Deborah Davis and Ezra Vogel. Cambridge, MA: Council on East Asian Studies, Harvard University, pp. 284–317.

———. 1991. "Birth planning between plan and market: The impact of reform on China's one-child policy," in *China's Economic Dilemmas in the 1990s: The Problems of Reforms, Modernization, and Interdependence*, ed. United States Congress, Joint Economic Committee. Washington, DC: United States Government Printing Office, pp. 252–269.

——— (guest ed.). 1992. "Family planning in China," *Chinese Sociology and Anthropology* 24, 2.

———. Forthcoming, 1994. *Rationing the Children: China's One-Child Campaign in the Countryside*. Berkeley: University of California Press.

Xie Zhenjiang. 1989. "There is no route of retreat," *Jingji Ribao (Economic Daily)*, 24 January 1989, in *Foreign Broadcast Information Service—Daily Report, China*, 15 February, p. 35.

Xinhua Domestic Service. 1986. "Plan of the CPC Central Committee and the State Council for Rural Work in 1986," 22 February, in *Foreign Broadcast Information Service—Daily Report, China*, 24 February, p. K6.

Xinhua News Service. 1991. "Communique on the 1990 Census," 19 June, in *Foreign Broadcast Information Service—Daily Report, China*, 24 June, pp. 39–41.

Yao Yilin. 1980. "Report on the arrangements for the National Economic Plans for 1980 and 1981," *Beijing Review* 23, 38 (22 September): 37.

Yu Youhai. 1980. "U.S. $1000 by the year 2000," *Beijing Review* 23, 43 (27 October): 16–18.

Zhao Ziyang. 1981. "The present economic situation and the principles for future economic construction," *Beijing Review* 24, 51: 29.

———. 1986. "Report on the Seventh Five-Year Plan," *Beijing Review* 29, 16: I–XX.

———. 1987. "Advance along the road of socialism with Chinese characteristics," *Beijing Review* 30, 45 (9–15 November 1987): I–XXVII.

Zhonggong guojia jihua shengyu weiyuanhui dangzu. 1984. "Report on the situation in family planning work," in *Selected Important Documents on Family Planning Since the Third Plenum of the Eleventh Central Committee* (in Chinese). Beijing: Central Party School Press, pp. 18–25.

Zhou Enlai. 1985. "Rural health work and family planning," in *China Population Almanac 1985* (in Chinese), ed. Population Research Center, Chinese Academy of Social Sciences. Beijing: China Social Sciences Press, p. 15.

Zweig, David. 1987. "Context and content in policy implementation: Household contracts and decollectivization, 1977–1983," in *Policy Implementation in Post-Mao China*, ed. David M. Lampton. Berkeley: University of California Press, pp. 255–283.

# Is Population Policy Necessary? Latin America and the Andean Countries

CARLOS ARAMBURÚ

FEW SCHOLARS WOULD DISPUTE the proposition that sustained improvements in the level of socioeconomic development will bring about a decline in fertility even in the absence of organized family planning programs. In the West, the demographic transition was accomplished not only without the benefit of family planning programs, but at a time when birth control was often considered immoral and when many governments were pronatalist in orientation (Glass, 1940). By 1965, prior to the introduction of significant family planning programs, fertility decline had been initiated in 42 less developed countries, many of them in Asia (Kirk, 1971). In Latin America, most countries have reduced their birth rates during the last 30 years despite only sporadic adoption of family planning policies prior to the 1980s (Nortman and Hofstatter, 1974; United Nations, 1987–90).

Demographic rates in Latin America display a complex relationship to antinatalist policy (Merrick, 1990). While Cuba, Panama, the Dominican Republic, Mexico, and Colombia, all of which adopted population policies in the 1960s and early 1970s, have reduced fertility between 40 and 60 percent from 1960–65 levels, El Salvador and Guatemala, whose population policies date from 1974 and 1975, have not achieved such success. Chile and Uruguay have almost reached replacement level without the benefit of such policy, and Argentina is not far behind. Moreover, several countries, including Venezuela, Brazil, Costa Rica, and Peru, have all reduced their fertility levels by more than 35 percent either without policies or with policies adopted only in the 1980s.

The demand factors that determine fertility are highly complex, and it would be naive to expect to find a strong relationship between fertility and any single development or policy factor. In Latin America, not only do the timing and pace of fertility decline seem to be weakly correlated with explicit fertility policies, but the relationship between policies and programs is also far from straightforward. Countries like Peru, with explicit population policies, often have weak official family planning programs and modest contraceptive prevalence rates, while others like Brazil, which lack a policy commitment to induce fertility decline, show high contraceptive prevalence (although a very limited method choice predominates).

To work backward, contraceptive prevalence is looked at first in terms of quantity (coverage) and quality (method mix), and its relationship to fertility levels. The issue of availability of contraceptives can then be assessed and linked to the types of family planning programs operating in each country. Finally, the issue of whether or not family planning effort is related to the presence of explicit population policies can be examined. A statistical analysis of program effort is not performed here, since this has been done convincingly by a number of authors (Mauldin and Lapham, 1985; Entwisle, 1989). Instead, the process by which policy, programs, and prevalence relate to one another in the three Andean countries, Ecuador, Bolivia, and Peru, is illustrated here.

## Fertility and contraception

Demographers have long agreed that the increase in contraceptive use is the most important proximate factor behind the reduction of fertility. Bongaarts (1984) showed that contraceptive prevalence could explain 85 percent of the variation in total fertility rates in a sample of 83 countries around 1980. Calculations (not shown) for the 19 Latin American countries listed in Table 1 show a negative correlation of –0.694 between total fertility (TFR) around 1989 and contraceptive prevalence (CPR) around 1985. This association increases to –0.712 when the TFRs are correlated with CPRs for modern methods.[1] Accordingly, it can be seen in the second and third columns of Table 1 that all countries with a TFR above 5 have "total CPRs" of around 30 percent while those countries with a TFR around 3 report that more than 50 percent of women of reproductive age use some form of contraception.

The gap between the total CPR and the CPR for modern methods can be used as a good indicator of the quality of contraception in each country; generally speaking, the higher the total CPR, the smaller the discrepancy between the two measures. Three countries show relatively large gaps between these two rates: Peru (23 percentage points), Paraguay (16 points), and Bolivia (15 points). Peru has a weak official family planning program and serious cultural barriers to the use of modern methods, despite a policy that favors fertility reduction. The same cultural barriers to the use of modern contraceptives seem to be present in Bolivia among the indigenous Aymara and Quechua populations. Information is lacking for Paraguay to confirm whether this is also true among the rural native population. Nei-

**TABLE 1 Latin America: Fertility and policy indicators**

| Region and country | TFR: Percent change 1965–85 | TFR 1989 | Contraceptive prevalence rate: Total/ modern methods 1985 | Government perception of fertility 1989 | Government policy on contraception 1989 | Government policy on population | Date |
|---|---|---|---|---|---|---|---|
| **Central America** | | | | | | | |
| Dominican Republic | –43.1 | 3.8 | 50/47 | High | A | A | 1968 |
| Nicaragua | –19.0 | 5.7 | 27/23 | High | A | No | na |
| Guatemala | –13.7 | 5.6 | 23/18 | High | A | B | 1975, 1984 |
| Honduras | –7.8 | 5.5 | 35/30 | High | A | B | 1982 |
| Mexico | –31.7 | 3.8 | 53/45 | High | A | A | 1974 |
| Costa Rica | –47.9 | 3.5 | 70/58 | High | A | B | 1982 |
| El Salvador | –13.9 | 4.4 | 47/45 | High | A | A | 1974, 1986 |
| Panama | –39.1 | 3.3 | 58/54 | Satisfactory | A | B | 1969 |
| Cuba | –50.6 | 1.8 | 79/— | Satisfactory | No | B | 1960 |
| **Temperate South America** | | | | | | | |
| Chile | –47.0 | 2.4 | 43/— | Satisfactory | A | No | na |
| Argentina | +7.3 | 3.1 | — | Satisfactory | B | No | na |
| Uruguay | +1.1 | 2.3 | — | Low | No | No | na |
| **Tropical South America** | | | | | | | |
| Ecuador | –27.5 | 4.3 | 44/36 | High | A | B | 1986 |
| Peru | –27.5 | 4.3 | 46/23 | High | A | A | 1985 |
| Bolivia | –7.4 | 5.5 | 26/11 | Low | B | B | 1989 |
| Colombia | –41.5 | 3.4 | 63/51 | Satisfactory | A | A | 1970 |
| Paraguay | –26.7 | 4.8 | 45/29 | Satisfactory | A | No | na |
| Venezuela | –36.5 | 3.4 | 49/38 | Satisfactory | A | No | na |
| Brazil | –38.0 | 3.4 | 65/56 | Satisfactory | A | No | na |

NOTES:
Column 5, A = Not limited, direct support; B = not limited, no government support.
Column 6, A = Demographic objective; B = nondemographic objective.
— = Data not available. na = not applicable.
SOURCES: For TFR Percent of change, author's estimates from Centro Latinoamericano de Demográfia, 1983. For TFR 1989 and CPR 1985, *World Population Data Sheet*, 1989. For government perception and policy on contraception, United Nations, 1989. For population policy and date, United Nations, 1987-90.

ther of these countries had official fertility policies at the time these data were collected, and most services were provided by the private and commercial sectors.

## Policy and programs in Latin America

According to the United Nations Population Fund (UNFPA), in 1985 the governments of 11 of the 19 Latin American countries listed regarded their fertility levels as satisfactory (United Nations, 1989). In the Dominican Republic, Mexico, El Salvador, Ecuador, and Peru, fertility was seen as too high, while the governments of Chile, Uruguay, and Bolivia viewed fertility as too low.[2] By 1989, the date shown in Table 1, the Bolivian government had changed its perception of fertility to too high, and stated that it placed no restrictions on the use of contraception; however, with population growth still regarded as too low, no direct support of contraception is offered to date by the government (United Nations, 1987–90). In fact, family planning programs are being cautiously offered at present in public hospitals, but only to avoid high-risk pregnancies.

Although until recently the majority of Latin American governments have expressed little concern for their countries' fertility levels, most of them have long supported access to contraceptive methods, either directly or indirectly (compare policy columns in Table 1). In most of these countries, contraceptive methods, especially barrier methods and oral contraceptives, are sold over the counter in pharmacies and small shops. The regulation requiring a prescription for oral contraceptives is largely ignored, and, in the cases of Bolivia and Peru, large, informal markets staffed mainly by female street vendors offer condoms, vaginal suppositories, and some oral contraceptives (usually donated) on the sidewalks and in marketplaces of cities and larger towns. The smuggling of contraceptives across the borders between Colombia, Ecuador, Peru, and Bolivia is also common, attributable to frequent fluctuations in exchange rates and large differentials in inflation levels. A growing demand for contraceptives has gone hand in hand with the agility of the informal sector in reaping profits and providing these commodities. While the quality of these informal services is questionable, by increasing the availability of contraceptive methods they undoubtedly contribute to the reduction of fertility. The size, mode of operation, and price structure of the growing informal market for contraceptives in Latin America require serious study so that the quality of these services can be improved.

Official population policy in Latin America is complex and still in flux; however, no policies were adopted until fertility had already started to decline. Of the nine Central American countries listed in Table 1, eight have official policies intended to influence fertility either for demographic reasons (listed as A in the "Population policy" column) or for health, human rights, or other nondemographic reasons (listed as B). Five of these countries adopted their policies in the 1960s or early 1970s, and some of them have reaffirmed their positions more recently (Nortman and Hofstatter, 1975; United Nations, 1987–90). Of the seven countries in tropical South America, only Peru has recently (1985) adopted an explicit antinatalist population policy. Colombia, one of the early adopters of a population policy (1970), no longer claims to have an explicit policy, but continues to provide support for family planning (United Nations, 1987). None of the three countries with low fertility levels in temperate South America has an antinatalist policy. Chile and Uruguay still express pronatalist sentiments, though Chile does provide some direct support for family planning. Argentina, by contrast, offers only indirect support (United Nations, 1987–90).

## Policy, programs, and prevalence in Ecuador, Peru, and Bolivia

The acceptance of family planning within a society is a consequence not only of the spread of contraceptive information and services, but more fundamentally of processes of social and economic change. Ideally, therefore, assessment of the effect of policies and programs would require that the cultural, social, and economic conditions in the countries under study be held constant. In Latin America, countries vary markedly both in their cultural traditions and in their levels of socioeconomic development. By focusing on the Andean countries, which, despite their differing socioeconomic levels, share a similar cultural tradition, some of these factors are controlled for in this chapter.

### Ecuador

In terms of socioeconomic development, Ecuador, with a 1989 population of around 11 million, and Peru, with more than 21 million, show some similarities (see Table 2). Ecuador clearly leads in terms of health status, while Peru has overtaken Ecuador in terms of gross national product (GNP) per capita, urbanization, women's education, and lower demographic growth. Compared with Ecuador, however, Peruvian society is less equitable and is troubled by social and political unrest. Bolivia, with a population of more than 7 million, is the least developed of the three Andean

**TABLE 2 Socioeconomic indicators: Ecuador, Peru, and Bolivia**

| Indicator | Ecuador | Peru | Bolivia |
|---|---|---|---|
| Total population 1989 (thousands) | 10,781 | 21,256 | 7,314 |
| Infant mortality rate (per 1,000 live births) (1985–90) | 63 | 88 | 110 |
| Life expectancy (years) (1988) | 66 | 62 | 53 |
| Population growth rate (percent) (1985–90) | 2.56 | 2.08 | 2.76 |
| Percent urban | 55 | 69 | 50 |
| Percent of women in secondary education (1987)[a] | 57 | 61 | 35 |
| Percent of workforce in agriculture (1986) | 39 | 40 | 46 |
| Gross national product per capita (1988, US$) | 1,120 | 1,300 | 570 |

[a]Percentage of relevant age group.
SOURCES: UN, 1990, Medium Variant; World Bank, 1986.

countries, showing high infant mortality, rapid population growth, and lower rates of urbanization, industrialization, and GNP per capita.

While Ecuador seems to be moving cautiously toward the formulation of an explicit population policy, albeit with nondemographic objectives (United Nations, 1987), there has been an active family planning program in the private, not-for-profit sector since 1966. In 1969, public services were instituted in the ministries of Public Health and Social Security, and in the armed forces. By 1973, however, family planning services covered a mere 1.8 percent of women of reproductive age.

The commercial sector has also been active in Ecuador, especially through a network of small shops characteristic of the smaller towns, where barrier methods and oral contraceptives are provided over the counter at prices comparable to those in the pharmacies. In 1982, a change in the law permitted sterilization for both medical and social reasons at the request of the client. As is shown in more detail later, legalization has led to an enormous increase in the prevalence of this method. In terms of program effort, Ecuador improved its program effort score by 10 percent or more according to the Lapham and Mauldin study covering the period 1972–82 (*Population Reports*, 1984: E142).

These measures have led to a contraceptive prevalence rate of more than 44 percent in 1987, up from 40 percent in 1982 (Ecuador, 1983; Centro de Estudios de Población y Paternidad Responsable, 1988 [Demographic and Health Survey]). Among married women, the proportion using modern methods has increased to 36 percent during the same period. In other words, more than 80 percent of married women practicing contraception are using a modern method. Associated with these developments, a fertility decline of 27.5 percent occurred between 1965 and 1985, the fastest drop being between 1965–75, when organized family planning programs were started. Ecuador can, therefore, be characterized as a country with moderate levels of contraceptive prevalence, good method mix, an active family planning program both in the private and public sectors, and a late start in terms of explicit fertility policy.

### Peru

Peru offers striking similarities to and differences from Ecuador's situation regarding family planning and contraception. Policy statements naming family planning as a basic social right were issued as early as 1976, but they were not translated into programs. In 1985, a full-blown population law was passed, and in 1987 a four-year population program was promulgated. In spite of this, the public sector program remains weak. The private, not-for-profit sector started limited-service programs in the capital of Lima in 1969, but all services were prohibited by the military regime in 1974 and were revived slowly after 1977.

Given the enormous economic inequalities and pervasive poverty in Peru, the private pharmaceutical market represents between 15 and 20 percent of total health expenditures. Sales in the industry have dropped sharply since 1980, reflecting the economic recession and the consequent decrease in the purchasing power of the majority of the population (Zschock, 1984). Furthermore, a combination of erratic government price regulations and increasing insecurity in the towns and cities of the highlands and in the poorest zones of the coastal urban centers has led to the closing of many pharmacies. Thus, the formal commercial contraceptive marketplace has shrunk and provides fewer contraceptives than are available in other countries of Latin America. A growing informal marketplace offering contraceptives at competitive prices seems to be taking over from the pharmacies, as donated commodities (especially condoms) find their way to the innumerable street vendors and market posts.

In this environment, total contraceptive prevalence among women in sexual unions increased by 10 percent between 1978 and 1981, but by less than 5 percent between 1981 and 1986, reaching 45.8 percent by that date (Peru, 1983; 1988). Despite a doubling of the proportion of married women using modern methods between 1981 and 1986, however, only half of them use efficient methods. Preliminary results from the 1991–92 Demographic and Health Survey (DHS) show that the contraceptive prevalence rate (CPR) has increased to 60 percent, but that traditional methods still account for more than 43 percent of total CPR (Peru DHS, 1992). Judging by the gap between safe and effective methods and total methods, method mix in Peru is the poorest in Latin America. This paucity of choice is the result of limitations in the availability of efficient and long-lasting methods as well as of cultural barriers that inhibit demand for modern methods despite Peruvians' overriding preference for smaller families.

In summary, Peru's situation can be described as one of intermediate contraceptive prevalence levels (similar to those in Ecuador), poor method mix, and weak public programs despite explicit population policies and plans aimed at reducing fertility.

## Bolivia

Bolivia represents perhaps the most difficult concentration of factors for the advancement of family planning and contraception. A series of governments, mostly military regimes, have maintained a strong pronatalist position since the 1960s, when the Peace Corps' alleged forced sterilization program was denounced through a widely publicized and acclaimed film, "La Sangre del Condor" (the blood of the condor). Among those who oppose any form of family planning, the position of the Catholic Church is perhaps the best articulated. In 1973, when the Ministry of Health requested UNFPA assistance in starting a nationwide maternal and child health and family planning program, Bolivia's Catholic bishops denounced the proposal, charging that "these programs, cloaked in an apparent altruistic desire to reduce the population so that people can reach a higher standard of living, cover selfish aims of international domination . . ." (quoted in Nortman and Hofstatter, 1975, p. 27).

In addition, close links exist between the Catholic Church and the only Bolivian organized labor movement, the nationwide Labor Union Movement. The Catholic Church, the labor movement, and the military, more than the political parties, constitute the key players in Bolivia's recent political history. Any political party or coalition that is elected to government has first to deal with these three forces if it is to survive. Much of the reluctance of party officials to address the family planning issue stems from fear of this trio rather than from a conviction that the population does not need or desire family planning. The official attitude is at odds with that of a majority of Bolivian women toward limiting their fertility. Thus, in 1983 a KAP (knowledge, attitude, and practice of family planning) survey in the ten largest cities of Bolivia showed that between 76 percent and 90 percent of women of fertile age agreed that family size should be limited (COBREH, 1985, Table 86, p. 113). Furthermore, between half and two-thirds of these urban women declared an ideal family size to be two or three children, while the TFR around that date was between four and six children for the same sample (p. 170).

That contraceptive prevalence levels in Bolivia are among the lowest in the region and have remained fairly constant since 1968 is not surprising, given these conditions. In 1989, only 30 percent of women in sexual unions were using a contraceptive method; but of these, 40 percent were using modern contraceptives (DHS, Bolivia, Table 5.5). This low level of modern-method use reveals not only low coverage, but also the poor quality of services, which tend to be limited to the better-educated and wealthier urban segments of the society. Thus, Bolivia stands out as one of the four or five countries in the region where fertility is still high (TFR 6.25 in 1985), showing only a slight decline since 1965.

Despite these serious political and institutional limitations, some progress has been made in the last several years. In 1987, through the sustained effort of a small group of Bolivian social scientists working in the National Population Council (CONAPO), the health authorities have begun to integrate family planning into the maternal and child health program. The goal is to improve the health of women and children and to avoid high-risk pregnancies. Individual physicians working in the Bolivian Social Security Service and in the Ministry of Health are responsible for this change of attitude. In addition, the new regime of president Paz Zamora, elected in 1989, is changing conservative views on family planning.

Recently, the private, not-for-profit sector has also been active in expanding education, information, and services. Nevertheless, these organizations are still weak, and their coverage is limited. An interesting strategy developed by some private voluntary organizations is to provide family planning services by agreement with local workers' organizations as an integral part of their health care. The recent economic crisis (1985–87), caused by a drastic decline in the price of tin, led to fiscal retrenchment and massive layoffs among workers in the state-owned mines, oil industry, and public services. In an attempt to reduce the social costs of basic support services, including family planning, they have been implemented since 1987 through a Social Emergency Fund. Bolivia's situation regarding family planning can, therefore, be described as lagging behind, but showing positive signs on the horizon.

## Sources of contraception

The relative weight of public, private, and commercial providers of contraceptive methods provides a valuable clue to the coverage and types of family planning programs in each country. For Ecuador and Peru, this information was obtained and published in the last round of DHS surveys. In Bolivia similar information is available for only a few urban centers from the KAP survey of 1983–84. The DHS data have some limitations for this analysis. First, the data are limited to one point in time, so that the process of change and the changing weights of each provider source cannot be estimated. Second, the question posed to the respondents asked for the source of the method they were using at the moment and/or the source of information they received on the method they are using. The data from responses to this question mix sources of supply and of information, making it impossible to differentiate between the two. Finally, the impact of the private, not-for-profit sector is probably underestimated since it is hard for the user to distinguish the institutional origin of health workers, who may be either health promoters from the public sector or community-based distribution workers from private organizations.

With these limitations in mind, the services for the three countries can be analyzed. In Table 3, the results for Ecuador around 1987 show an almost even split between the public sector, supplying approximately 38 percent of the methods and the private, for-profit sector, which served around 40 percent of all contraceptive users. The private, nonprofit sector supplied roughly 15 percent of users. Female sterilization, the most commonly used method, is provided mainly by the public sector (for 58 percent of all users) and in particular by the Ministry of Health (for 55 percent of all users of this method); private, for-profit physicians and clinics serve around 35 percent of all sterilized women, while private, not-for-profit organizations perform only 6 percent of sterilizations. By contrast, the private, for-profit sector seems to provide most of the short-term methods: 52 percent of oral contraceptives, 85 percent of injectables, and 53 percent of

**TABLE 3 Percent distribution of women who use contraceptives by source of supply, according to method, Ecuador, Peru, and Bolivia**

| Ecuador | | | | | | | |
|---|---|---|---|---|---|---|---|
| **Source of supply** | **Pill** | **IUD** | **Inject-ables** | **Barriers[b]** | **Steril-ization** | **Rhythm** | **All methods** |
| **Public sector** | **33.0** | **28.0** | **10.0** | **29.0** | **58.0** | **15.0** | **38.0** |
| Ministry of Health | 31.0 | 25.0 | 10.0 | 29.0 | 55.0 | 15.0 | 36.0 |
| Social Security | 1.0 | 2.0 | 0.0 | 0.0 | 2.0 | 0.0 | 1.0 |
| Armed Forces | 1.0 | 1.0 | 0.0 | 0.0 | 1.0 | 0.0 | 1.0 |
| **Private, for-profit** | **52.0** | **34.0** | **85.0** | **53.0** | **35.0** | **35.0** | **40.0** |
| Pharmacy | 20.0 | 0.0 | 15.0 | 21.0 | 0.0 | 0.0 | 6.0 |
| Physician/clinic | 32.0 | 34.0 | 70.0 | 32.0 | 35.0 | 35.0 | 34.0 |
| **Private, nonprofit** | **15.0** | **38.0** | **5.0** | **16.0** | **6.0** | **7.0** | **15.0** |
| **Other[a]** | **0.0** | **0.0** | **0.0** | **2.0** | **1.0** | **43.0** | **7.0** |
| Total (N=1,283) | 100.0 | 100.0 | 100.0 | 100.0 | 100.0 | 100.0 | 100.0 |
| Percent of women using method | 20.4 | 22.8 | 2.3 | 3.0 | 36.9 | 14.5 | 100.0 |

[a] Includes relatives, friends, and informal sources (for example, street vendors).
[b] Includes condoms and vaginal suppositories and foams.
SOURCE: Centro de Estudios de Población y Paternidad Responsable, 1988, Table 4.9

| Peru | | | | | | | |
|---|---|---|---|---|---|---|---|
| **Source of supply** | **Pill** | **IUD** | **Inject-ables** | **Barriers[b]** | **Steril-ization** | **Rhythm** | **All methods** |
| **Public sector** | **57.7** | **39.6** | **9.1** | **30.4** | **68.1** | **32.8** | **49.0** |
| Ministry of Health | 46.4 | 30.2 | 6.1 | 21.7 | 51.9 | 26.3 | 40.5 |
| Other public hospital | 10.8 | 4.7 | 3.0 | 8.7 | 16.2 | 2.6 | 6.0 |
| Health worker | 0.5 | 4.7 | 0.0 | 0.0 | 0.0 | 3.9 | 2.5 |
| **Private, for-profit** | **32.4** | **48.8** | **33.3** | **17.4** | **29.2** | **13.4** | **36.5** |
| Pharmacy | 0.0 | 9.3 | 9.1 | 0.0 | 0.0 | 0.4 | 4.5 |
| Physician/clinic | 32.4 | 39.5 | 24.2 | 17.4 | 29.2 | 13.0 | 32.0 |
| **Private, nonprofit** | **3.6** | **2.3** | **6.0** | **4.4** | **0.0** | **2.8** | **2.5** |
| **Other[a]** | **6.3** | **9.3** | **51.6** | **43.5** | **2.7** | **46.2** | **9.0** |
| **Source not reported** | **0.0** | **0.0** | **0.0** | **4.3** | **0.0** | **4.8** | **3.0** |
| Total (N=1,045) | 100.0 | 100.0 | 100.0 | 100.0 | 100.0 | 100.0 | 100.0 |
| Percent of women using method | 21.2 | 4.1 | 3.2 | 2.2 | 17.7 | 51.6 | 100.0 |

[a] Includes relatives, friends, and informal sources (for example, street vendors).
[b] Includes condoms.
SOURCE: Peru, 1988.

barrier methods (here including condoms, vaginal suppositories, and contraceptive foams). Private physicians and clinics are reported as the main source of supply for all these methods, rather than pharmacies. Finally, the private, not-for-profit sector seems to provide most of the IUDs in Ecuador (38 percent). The inclusion of the rhythm method in the data poses a question because it refers to the source of information rather than supply (since this method cannot be supplied). Most users of this method (43 percent)

**TABLE 3 (continued)**

| | Bolivia[a] (city of La Paz) | | | | | | |
|---|---|---|---|---|---|---|---|
| Source of supply | Pill | IUD | Inject-ables | Barriers[b] | Steril-ization | Abortion | All methods |
| **Public sector[c]** | **2.0** | **6.6** | **2.2** | **2.5** | **10.3** | **6.5** | **5.0** |
| **Private** | **97.2** | **91.8** | **97.8** | **96.7** | **89.7** | **91.5** | **94.1** |
| Pharmacy | 68.3 | 3.1 | 38.4 | 71.0 | 3.4 | 1.2 | 31.1 |
| Physician | 27.7 | 11.5 | 57.6 | 24.6 | 66.8 | 66.3 | 42.1 |
| Clinic | 1.2 | 77.3 | 1.8 | 1.1 | 19.5 | 24.0 | 20.8 |
| **Other** | **0.8** | **1.6** | **0.0** | **0.8** | **0.0** | **2.0** | **0.9** |
| Total (N=1,532) | 100.0 | 100.0 | 100.0 | 100.0 | 100.0 | 100.0 | 100.0 |
| Percent of women of reproductive age who ever used method | 6.3 | 7.3 | 1.1 | 5.2 | 1.1 | 1.7 | 22.7 |

[a] All women of reproductive age who knew a method were asked where they would go to get it.
[b] Includes condoms, vaginal suppositories, and other barriers.
[c] Ministry of Health.
SOURCE: COBREH-Pathfinder, 1985, Tables 59 and 125.

quote friends, relatives, or other nonprofessional sources as providers of information.

Such costly, permanent methods as sterilization and IUDs are provided by subsidized sources including the Ministry of Health and the not-for-profit organizations, while the temporary, less expensive contraceptives are obtained mainly from private clinics or physicians and pharmacies.

Comparable data for Peru are also presented in Table 3. For all methods, the public sector, especially the Ministry of Health, is the main provider of contraceptive services, covering 49 percent of active users. Second is the private, for-profit sector with 36.5 percent and especially, as in Ecuador, private clinics and physicians (32 percent). The private, not-for-profit sector has a small share of the market, 2.5 percent of users of all methods. "Other sources" account for 9 percent of all methods, particularly injectables (52 percent), barrier methods (condoms) (43 percent), and rhythm (46 percent). These figures represent a variety of sources of supply and information, but they point to the importance of the informal sector as a provider of these methods either through health practitioners (for contraceptive injections) or street vendors (for condoms). As in Ecuador, female sterilization is provided mainly by the public sector, which serves 68 percent of all users of this method, but in Peru the public sector also provides most of the oral contraceptives (58 percent). The private, for-profit sector provides most of the IUDs (almost 49 percent) and about one-third of pills, injectables, and female sterilizations.

Two characteristics stand out in Peru: the weight of the informal sector as a provider of both information and barrier methods, and the relative weakness of the private, not-for-profit sector, despite the large numbers of private voluntary organizations operating in the country.

Finally, the data for Bolivia are presented in Table 3. An important difference between Bolivia and the other two countries is that the information refers to all women who declared knowledge of at least one method in response to a question posed in hypothetical form: "To whom would you go to get this method?" The results are revealing: More than 94 percent of users of all methods rely on the private sector for their contraceptives, mainly on physicians, pharmacies, and clinics. The public sector supplies only 5 percent of contraceptive users. The data do not distinguish between private, for-profit clinics and those associated with voluntary organizations, so it is not possible to assess the relative importance of these sources. Within the private sector, however, pharmacies supply most of the temporary methods (pills, 68 percent; barriers, 71 percent) while physicians provide most of the injectables (58 percent), sterilizations (67 percent), and abortions (66 percent).

Given the unfavorable environment for family planning in Bolivia, the virtual nonexistence of public programs, and the limited coverage of the private voluntary effort, the private commercial sector appears to be providing most contraceptive services. These sources tend to be concentrated in the larger urban centers; they charge higher fees than other sources and are not actively involved in providing information and counseling about the health and other benefits of family planning. Only the better-educated, wealthier urban groups have access to family planning services. This circumstance is unfortunate, given the large potential demand for limiting family size as expressed by most Bolivian women and the high incidence of abortion (20 out of every thousand women of reproductive age undergo abortion in Bolivia, although it is illegal) (Centro de Orientación Familiar, 1987, p. 82).

Although the public sector is often regarded as the main or even the only player in the family planning effort in Latin America, the reality is far more complex, and other actors must be taken into account. Insight about family planning can be derived from a breakdown of private-sector providers as to whether or not they operate for profit. The informal sector seems to play a major role by providing barrier methods and contraceptive pills. This source of supply merits careful study, as little is known about it.

## From policy to program: Peru's population policy

The passage of the National Policy on Population on 5 July 1985 (Decree Law No. 346), a few days before the first democratic transfer of power in the history of the Peruvian republic, represented the culmination of more than 20 years of effort to address the country's population problems. The first sign of official concern had appeared in 1964 with the creation of the Center for Population and Development Studies. In 1968, however, the military government prohibited all public and private family planning activities, closing the few isolated clinics that had started to offer services. At the World Population Conference in Bucharest in 1974, the military government maintained its pronatalist stance, but two years later modified its position and issued a document entitled "Guidelines on Population Policy." Although the document had a progressive cast, emphasizing the importance of human dignity, the guidelines were never incorporated into any official policy. Indeed, official attitudes toward population at this time were imbued with a sense of optimism based on the expectation that structural change would be sufficient to resolve any problems caused by rapid population growth.

A major economic crisis, starting in 1975 and continuing to this date, changed official views on rapid population growth. Gradually the impression grew that the slow growth of production and services, rapid and disorganized urban expansion, and the precarious levels of health and living standards of a vast majority of the population were exacerbated by population growth. The previous decade had seen a burgeoning of research and training on population issues in universities and in public and private research institutions. The growing number of individuals who were aware of demographic issues came together in such private organizations as the Asociación Multidisciplinaria de Investigación y Docencia en Población (AMIDEP), and were gradually able to influence key opinion groups, sensitizing them to Peru's population problem. The impact was widely spread throughout the society. By the 1979 elections, for example, both the party of the liberal right, the Partido Popular Cristiano, and the Marxist left party, the Izquierda Unida, espoused plans for dealing with population questions. The Peruvian Constitution, proclaimed in 1979, stated in Article VI that "the State favors responsible parenthood."

The first practical steps toward the elaboration of a population policy were taken in 1980, under the new democratic regime of President Fernando Belaunde, with the creation of the National Population Council (CNP) within the office of the Prime Minister.[3] Despite the goodwill of its first leaders, the policymaking character of the CNP limited its active role in implementing family planning programs, which were assigned primarily to the Health and Education Sectors. The immense bureaucratic inertia of these institutions in the face of a new and controversial issue determined the slow progress of the national family planning program. Additionally, the implementing agencies lacked the political support of the highest authorities, including the president, that might have stimulated more vigorous implementation of the policy.

The preparation and approval by Congress of the Peruvian Population Law was a slow process that occupied the entire five years, 1980–85, of the regime of the Acción Popular. The first step, achieved through the exertions of the president of the CNP, was the establishment in January 1985 of a nonpartisan parliamentary commission charged with preparing the draft of the Population Law. In addition to ten representatives from the Senate and lower chamber of Parliament, the commission included five representatives of the executive branch drawn from the ministries of Health, Education, and Justice, the Planning Institute, and the CNP. Once installed, the commission was provided with an advisory task force comprising ten technical and professional officials drawn from appropriate bureaucratic agencies and the private sector. By mid-April, the task force had prepared a draft for presentation to the parliamentary commission. This draft was reviewed, debated, and modified in the course of 19 working sessions, and received final approval by parliament on 14 June. On 5 July, as noted earlier, President Belaunde proclaimed the Population Law of Peru.

As was the case in the 1979 elections, members of the commission approached their work in a nonpartisan spirit. For example, representatives of the liberal right PPC, the Marxist, and populist parties were among those who supported an aggressive policy of population control, while more conservative positions were supported by a PPC senator and by representatives of the ruling party and the executive branch. The final version of the law reflects a compromise with other compelling interests. In particular, the law reflects the pressure exerted by the Catholic Church to exclude sterilization and abortion as family planning methods. Article VI, which specifically excludes these two methods, was inserted into the final version as a result of conversations between the Minister of Justice and representatives of the Church. The commission was motivated to include Article VI in order to avoid a direct confrontation with the Episcopal Commission, which might have resulted in a public argument with the powerful Catholic hierarchy. One of the few interest groups that publicly opposed the policy was a right-wing, pro-life association that was and is closely associated with some of the more conservative elements of the Church. In a series of articles in the press, this organization denounced what it termed "the genocidal attempt of the government to sterilize Peruvian women" (*El Comercio*, 12 July 1985).

The Population Law was approved with little public opposition or debate and was passed almost unnoticed. External circumstances including the increased activities of the Marxist terrorist group "Shining Path," high inflation, and especially the imminent change of government after the April general elections, obscured the population legislation. This lack of public awareness may have determined that during the first year of President Alan García's mandate, the implementation of the new law was insignificant. In

November 1986, President Alan García, addressing private entrepreneurs and businessmen, first mentioned rapid population growth as a major obstacle to national development, and only then was a concrete population program prepared under his direct supervision and passed as law in June 1987.

### Design and implementation of the program

The National Population Program was designed over three years, 1987–90, by a broadly based commission comprising representatives of the Ministries of Health and Education, the Planning Institute, the National Population Council, and the Statistical Institute, as well as the Church, the private sector, and two personal delegates of the president. At the commission's opening session, President García asserted that he regarded rapid population growth as a major obstacle to socioeconomic development, one that impeded the progress of social justice for poor families. The president also stated that fertility levels should be reduced through the widespread provision of modern contraceptive methods, including sterilization. He was concerned about reaching the rural and urban poor and stressed the need to include men within the scope of the program. Finally, the president expressed his intention to confront the Church personally if necessary, and suggested that a long-term national consensus should be developed on the need to address population issues.

Like most contemporary population policies, the program developed by the commission in Peru has broad objectives, including an intention to reduce morbidity and mortality rates as well as to modify the distribution of the population. In terms of fertility, the objective is to reduce the birth rate in order to attain a better balance between population, development, and resources. The program is also intended to create a social and cultural environment favorable to responsible parenthood through the dissemination of information and education. In pursuit of the fertility objectives, the program established quantitative targets, which include reducing the TFR from 4.4 in 1987 to 3.7 by 1990, thus bringing about a corresponding decline in population growth from 2.6 percent to 2.2 percent in the same period. To achieve these goals, contraceptive prevalence of modern methods would have to increase from 28 percent in 1986 to 32 percent by 1990. In fact, according to the 1991–92 DHS, the TFR was 3.5 by 1990 and CPR for modern methods rose to 33 percent. The goal was met despite the weak nature of the public sector family planning program.

In July 1990, Alberto Fujimori was elected president. A few months later, he took a clear position in favor of family planning, eliciting a strong reaction from the Catholic hierarchy. The president responded aggressively, denouncing the Church's "medieval and reactionary attitudes." He declared

1991 as the year of "Austerity and Family Planning." Despite this explicit support and the drafting of a Population Program for 1990–95 by the National Population Council, implementation of the family planning program faced serious constraints. Two sets of difficulties arose, one related to the structure and culture of the political administrative system, the other reflecting specific problems in the program's design.

The program was conceived as a coordinated effort involving a number of government agencies. This conception was at odds with the sectoral character of public administration in Peru. Responsibility for family planning services, information and education, research and evaluation, the improvement of women's status, and other program components was assigned to separate autonomous agencies that were unwilling or unable to coordinate their activities. Family planning services were offered by three agencies, the Ministries of Health and Social Security and the health services of the Armed Forces, while the National Population Council, the Statistical Institute, and the Planning Office of the Ministry of Health all had responsibilities for research and evaluation. Finally, the National Population Council, which was charged with the overall coordination of the policy, lacked the authority and political leadership for the task, having been placed at a subordinate level in the Ministry of Health. As a consequence, each public agency developed its own activities independently of the others, weakening the overall effort and diminishing its social impact.

Internal divisions in the bureaucratic structure and culture of Peru's administrative agencies created further difficulties. Service ministries like those of Health and Education have three distinct categories of personnel, each with different training, responsibilities, and terms of tenure. Perceptions of the program vary at different levels of the hierarchy.

At the highest levels of the bureaucracy, executive positions are political and appointments are made by each new regime when it takes office. Incumbents of these positions seldom have technical or administrative expertise, and their tenure is usually short. Because of their party affiliation, political appointees are sensitive to new program directions and innovative policies, but their functions within the agency isolate them from daily service delivery. Next in the hierarchy are career bureaucrats, technically trained, well versed in the intricacies of public administration, and perceiving themselves as politically neutral. These are the medical service personnel in hospitals and health-care facilities and also the rank and file of the powerful unions of public workers. For most people in these positions, new programs and activities are an additional burden and are viewed as disruptive of familiar routines. In the absence of incentives for assuming additional responsibilities, career bureaucrats often respond passively when new actions are called for. The third level, the field staff, is usually recruited from the ranks of young professionals and technicians. Members of this cadre frequently obtain their positions by exploiting their kinship or other personal ties with members of the political bureaucracy. This group is commonly motivated by the desire to gain experience and to move on to the private sector. Turnover is high, in part because of low salaries and the dampening influence of middle-level superiors.

Among the second group of difficulties encountered by the new program, four potentially remediable problems stand out: the strictures on financial resources available to the program, the lack of an incentive system for program staff, the absence of an efficient logistics system for the supply of contraceptives, and the static mode of service delivery.

*Financial resources* The program in the Ministry of Health relies heavily on international donor funds. Outside funding is tied to specific projects that are centrally controlled and only marginally integrated into regular ministry services. In 1988, for example, only one-fifth of program expenditures were covered by national funds. Few resources were available outside the main urban areas or trickled down to health clinics or smaller hospitals. In addition, family planning was perceived by other health programs as a wealthy department because of its outside funding, and was subjected to constant pressures for resource sharing and for positions within the program.

*Incentive system* Since 1986, the family planning program has been organized as an integral part of maternal and child health services under the direction of the head of obstetrics and gynecology in each hospital. No incentive payments were awarded for additional responsibilities, nor were the departments given supplementary resources other than contraceptives. The additional burden was placed on the maternal and child health departments at a time when health workers increasingly saw themselves as underpaid and overworked. During the early years of the family planning program, strikes among health workers became common, resulting in the closure of nonemergency services, sometimes for months at a time.

*Logistics and programming services* The program suffered from the failure to develop a decentralized and reliable logistics system. As a result, some areas lacked contraceptives, while others received too many. Neither program evaluation nor target setting with specific coverage goals became institutionalized. Insufficient resources were available for supervision and for the installation of monitoring systems, and little was done to train program directors in these techniques. Overall, therefore, the quality of services remained poor.

*Static services* Notwithstanding Peru's verbal commitment to strengthening its primary health-care system, services in the Ministry of Health remain almost exclusively hospital based, with little development of public health services at the community level. Consistent with this pattern, the family planning program did not develop a system of community-based distribution of contraceptives or of information and education. This dearth of outreach is particularly unfortunate because there is a widespread reluctance to visit a hospital if one does not feel ill. Long waiting periods in the clinics, unfriendly staff, and poor counseling leave clients' fears of side effects unaddressed. This failure is a serious deterrent to continued contraceptive use, as surveys show that fear of side effects is the primary barrier to acceptance of modern methods, especially of hormonal methods and IUDs, among women who say they do not want any more children (Aramburú and Li, 1989).

To suggest that services are uniformly bad would be incorrect and unfair. In some hospitals and clinics, personnel are making valiant efforts to increase coverage and improve services. Their isolated attempts, however, do not add up to an efficient or effective national system of service delivery in family planning.

## Conclusion

This chapter has examined the complex and indirect relationships between fertility decline, contraceptive prevalence, population policy, and family planning programs in Latin America. The countries of the region differ widely in respect of these relationships as well as in the socioeconomic and cultural conditions that influence reproductive decisionmaking. Nevertheless, the analysis supports the following conclusions.

First, despite differences in timing and pace, fertility decline has been widespread throughout the region since the 1960s. Generally speaking, the decline started approximately ten years before any explicit form of population policy was implemented. In nine of the more developed countries, fertility declined by more than one-third between 1965 and 1985 with or without a fertility-reduction policy. Moreover, the size of the decline is unrelated to the level of fertility at the beginning of the process.

Second, fertility is significantly and negatively correlated with contraceptive prevalence and with the use of modern methods. High prevalence rates also indicate that a large proportion of women are using efficient methods. The large gap between total prevalence and modern method prevalence found in Peru, Bolivia, and Paraguay suggests that these differences may be indicative of weak programs in those countries.

Third, in the three Andean countries examined in greater detail, the presence of an official fertility reduction policy appears to be neither a necessary nor a sufficient condition for the emergence of a strong family planning program. Ecuador, which is still moving toward the adoption of such a policy, has a stronger program in terms of coverage and quality than Peru, which had an earlier and stronger commitment to induced fertility decline.

Bolivia, lacking both a population policy and an official family planning program, shows both low prevalence rates and poor quality of method mix.

Finally, however, the analysis shows that explicit population policies can help to reduce the unevenness of the transition period, provided they are accompanied by strong, well-organized family planning programs. Under such circumstances, a population policy can also lower the social costs of the program by making available safe and effective contraceptive methods. Even where government programs are weak, a population policy can create a favorable environment for the expansion of private family planning efforts and donor assistance. A population policy, therefore, while not essential in bringing about a fertility decline, can do much to ease the pain of the demographic transition.

## Notes

1 These associations are both highly significant statistically (F=.001).

2 By 1989, Nicaragua, Guatemala, Honduras, and Costa Rica had changed their assessments from satisfactory to too high. Chile had also changed its assessment, from too low to satisfactory (United Nations, 1989).

3 The author acted as Planning and Evaluation Director of the CNP from 1981 to 1985. He was a member of the task force that drafted the Population Law in 1985, and later participated in the Presidential Commission that prepared the first Population Program (1986–90) under President Garcia's regime. Between 1990 and 1991 he was President of the National Population Council under President Fujimori.

## References

Aramburú, Carlos, and Dina Li. 1989. "La Anticoncepción en Cinco Ciudades del Perú: Mito y Realidad," *Lucar de Encuentro*, Año 3, No. 1–2, January–April.

Bolivia. 1990. *Encuesta Nacional de Demografia y Salud, 1989.* La Paz, Bolivia: Instituto Nacional de Estadística.

Bongaarts, John. 1984. "Implications of future fertility trends for contraceptive practice," *Population and Development Review* 10, 2: 341–352.

Centro Latino Americano de Demographia (CELADE). 1983. *Boletin Demográfico* no. 32. Santiago, Chile: CELADE.

Centro de Estudios de Población y Paternidad Responsable (CEPAR). 1988. *Encuesta Demográfica y de Salud Familiar, 1987.* Quito, Ecuador: CEPAR.

Centro de Orientacion Familiar (COF). 1987. *Bolivia: Aspectos Demográficos*. La Paz, Bolivia: COF.

COBREH–Pathfinder (Consultora Boliviana de Reproduccion Humana). 1985. *Influencia de los Aspectos Socio-Economicos y Culturales en la Reproducción Humana*. La Paz, Bolivia: COBREH.

Consejo Nacional de Población (CONAPO). 1989. "Mujer, trabajo y Reproducción humana." La Paz, Bolivia: CONAPO–Pathfinder.

Ecuador. 1983: *Encuesta de Salud Materno—Impanial y Variables Demográficas, 1982.* Quito, Ecuador: ININMS.

*El Comercio*. 12 July 1985, Lima, Peru, p. 12.

Entwisle, Barbara. 1989. "Measuring components of family planning program effort," *Demography* 26, 1: 53–76.

Freedman, Ronald, and Deborah Freedman. 1989. "The role of family planning programs as a fertility determinant." Paper presented at the seminar on the Role of Family Planning Programs as a Fertility Determinant, sponsored by the IUSSP Committee on Comparative Analysis of Fertility and Family Planning. Tunis, 26–30 June.

Glass, David V. 1940. *Population Policies and Movements in Europe*. Oxford: Clarendon Press.

Kirk, Dudley. 1971. "A new demographic transition?" in *Rapid Population Growth: Consequences and Policy Implications,* Vol. II, ed. National Academy of Sciences. Baltimore, MD: Johns Hopkins Press, pp. 123–147.

Mauldin, W. Parker, and Robert J. Lapham. 1985. "Measuring family planning program effort in developing countries, 1972–82," in *The Effects of Family Planning Programs on Fertility in the Developing World,* ed. Nancy Birdsall. Working Papers, no. 677. Washington, DC: The World Bank.

Merrick, Thomas. 1990. "The evolution and impact of policies on fertility and family planning: Brazil, Colombia, and Mexico," in *Population Policy: Contemporary Issues,* ed. Godfrey Roberts. New York: Praeger, pp. 147–166.

Nortman, Dorothy, and Ellen Hofstatter. 1974. *Population and Family Planning Programs: A Fact Book.* New York: The Population Council.

Peru. 1979. *Constitucion de la Republica,* Lima, Peru.

———. 1983. *Aspectos Demográficos y Prevalencia de Anticonceptivos en el Perú: Resultados de la lera Rencuesta Nacional de Prevalencia de Anticonceptivos.* Lima, Peru: Instituto Nacional de Estadistica, Ministerio de Salud y Westinghouse Health Systems.

———, Instituto Nacional de Estadistica. 1988. *Encuesta Demográfica y de Salud Familiar, 1986.* Lima, Peru: Informe General.

———. 1992. *Encuesta Demográfica y de Salud Familiar 1991–92.* Lima, Peru: Instituto Nacional de Estadística e Informática, Asociación Benéfica Prisma, DHS International.

*Population Reports.* 1984. "Law and policies affecting fertility: A decade of change." Series E, no. 7. Baltimore, MD: Population Information Program, Johns Hopkins University.

United Nations. 1987–90. *World Population Policies,* vols. 1–3. New York: United Nations.

———. 1989. *Global Population Data Base.* New York: United Nations.

———. 1990. *World Population Prospects.* New York: United Nations.

World Bank. 1986. *World Development Report 1986.* New York: Oxford University Press.

*World Population Data Sheet.* 1989. Washington, DC: Population Reference Bureau.

Zschok, D. 1984. "Medical care under social insurance in third world countries," in *Alternative Health Delivery Systems: Can They Serve the Public Interest in Third World Settings?* ed. National Council for International Health. Occasional Papers, Washington, DC: NCIH, pp. 17–23.

# The Politics of Research on Fertility Control

DONALD P. WARWICK

POLITICS AND RESEARCH on fertility control shape each other. Research affects fertility control by showing the harms or benefits of population growth, the presence or absence of popular interest in family planning, and the effectiveness or futility of programs designed to change the numbers of children born. Politics bears on research by affecting every critical decision about what is to be done, from the need for a study to the final uses of the data. These transactions take place among scholars favorable, opposed, and neutral with respect to fertility control.

For advocates of change, the beauty of population research is that, under the auspices of science, it moves toward action. Frank Notestein, former president of the Population Council, commented:

> Probably the best way to make progress in a dangerous field is to sponsor "research" rather than "action." Who can be against "truth"? Study the field, and let everyone come to his own conclusions, the slogan can be. Keep clear of advocacy. So research becomes a substitute for action. . . . With all its delay, I am inclined to think that it is also the best way to foster "action." (1982: 684)

In population studies, the most common example of politically productive research was the Knowledge-Attitudes-Practice (KAP) survey. Carried out in dozens of countries, KAP surveys nominally assessed what fertile individuals knew, felt, and were doing about birth control. But KAP studies also quelled the fears of national leaders about starting population control programs. According to Bernard Berelson,

> Such a survey should probably be done at the outset of any national program—partly for its evaluational use, but also for its political use, in demonstrating to the elite that the people themselves strongly support the program and in demonstrating to the society at large that family planning is generally approved. Within not too large limits, the major results of such surveys can

> now be pretty well predicted, but they are still highly useful for their persuasive and informational impact. (1964: 11)

For Joseph Stycos, KAP surveys were not only a technique of persuasion, but also an avenue to action.

> [R]esearch is a relatively uncontroversial way of initiating activity in population control, in countries where direct efforts are not possible. The research itself, in addition to providing valuable information for possible future programs, stimulates the interest of those directly and indirectly involved, and may serve to accelerate the whole process of policy formation. (1964: 368)

The main scientific pitfall in politically useful research is the distortion of design, data collection, and interpretation. Paul Demeny is skeptical about the objectivity of much research on population policy.

> [S]ocial science research directed to the developing countries in the field of population has now become almost exclusively harnessed to serve the narrowly conceived short-term interests of programs that embody the existing orthodoxy in international population policy. . . . Invoking the supposed urgency of the problems it is trying to solve, the population industry professes no interest in social science research that may bear fruit, if at all, in the relatively remote future. Equally, it disdains work that may be critical of existing programs, or research that seeks to explore alternatives to received policy approaches. It seeks, and with the power of the purse enforces, predictability, control, and subservience. (1988: 470–471)

Julian Simon is harsher, labeling as "corruption . . . the nexus of connections among research funding, individuals' perquisites, individual and institutional decisions about research topics to pursue, choices of people to hire and invite, emphasis placed upon various findings in the research, and sometimes the research conclusions themselves" (1990: 39–40). Others, such as Ronald Freedman (1987), emphasize the positive contributions of social science research to population policy and assess its quality as high.

This chapter explores the impact of politics on population research and of population research on politics. It poses two central questions: (1) How does politics affect the accuracy and honesty of research on fertility control? and (2) What are the political uses of studies on fertility control? The primary focus is on the politics–research nexus in the United States.

## The political setting

The main sources of political influence on population research are international donors and other funding organizations; domestic politics; local organizations involved with fertility control; the researchers themselves; and the organizations for which they work.

### Funding agencies

The major agencies supporting research on fertility control are officially committed to limiting population growth. The leading contributor of funds for such research is the United States Agency for International Development (USAID). Other major donors include the governments of Canada, Denmark, Finland, Germany, Japan, the Netherlands, Norway, Sweden, and the United Kingdom (United Nations Population Fund, 1992). The Mellon Foundation, the Ford Foundation, the Rockefeller Foundation, and the Hewlett Foundation contribute private funds to support fertility control.

In addition to USAID, the agencies most heavily involved with research on birth limitation are the United Nations Population Fund (UNFPA) and the World Bank (Sadik, 1990; Sai and Chester, 1990). The main nongovernmental agencies include the International Planned Parenthood Federation; the Population Council; the Association for Voluntary Surgical Contraception; Family Planning International Assistance, the international arm of the Planned Parenthood Federation of America; the Pathfinder Fund; Family Health International; and Population Action International, formerly the Population Crisis Committee.

How do these organizations interact with each other, and what difference do the interactions make for research? Numerous studies have shown the financial, personal, political, and intellectual connections among the leading funding sources (Bachrach and Bergman, 1973; Piotrow, 1973; McCoy, 1974; Kasun, 1988; Hartmann, 1987). Smaller organizations, such as the Pathfinder Fund and the Population Council, are financially dependent upon the large donors, especially upon USAID and UNFPA. Many of these agencies also share common values and work on joint projects. Bachrach and Bergman (1973: 77) describe a "Population Coalition" whose members call it a "symbiotic system," a "Mafia-like structure," and an "interlocking directorate." The Pathfinder Fund (1990) gave specific examples of how nongovernmental population organizations work together on matters ranging from research to lawsuits over government restrictions on funding for abortion.

The historical pattern was that small private organizations took initiatives that encouraged action by larger institutions, such as USAID. The success of the foundations then stimulated movement by governments. They, in turn, pushed for support by the United Nations and other international donors.

> The situation is almost like the links of a food chain. The personally led special-purpose foundations experiment for and nourish the larger and more

> deeply institutionalized foundations and universities. These, in turn, experiment for and nourish the governments, which now show signs of experimenting for and nourishing the international organizations. The same activity that is viewed as improper, if not downright wicked, at the beginning of the chain is transformed by the end into an essential constituent of virtue if not a basic human right. (Notestein, 1971: 82)

Once governments and international organizations entered the scene, they sponsored research on fertility control and used the resulting knowledge to design and improve national family planning programs.

### Domestic politics

Whether population research can be carried out in a given country depends on the domestic politics of fertility control. Political dynamics affect the very possibility of doing research on population. In some countries, such as Peru in the early 1990s, the topic of birth control was so sensitive that researchers would have taken major risks in doing village interviews on that subject. Other gatekeepers are the national organizations responsible for population policy and family planning programs. They have their own ideas of what research should be done, what findings they want to see, and whether the results should be published.

### Researchers and their organizations

Self-selection takes place in all fields of intellectual inquiry, including fertility control. To judge from the literature, most researchers in this field approve of family planning programs or of placing socially enforced limits on population growth. These preferences affect the topics they choose for study, their research methodology, and their interpretations of findings. The same is true of those, such as Mamdani (1972), Kasun (1988), and Aird (1990), who find fault with fertility control programs. The critics are as unlikely to carry out KAP surveys as advocates are to report coercive pressures on clients.

Also vital to politics is the organization that employs population researchers. Research institutions do not control all the thoughts and writings of their staff, but there are points where institutional interests and scholarly freedom do clash. One is when staff publications criticize the ideology, policies, and actions of the organization's donors. Managers can set limits on writers' freedom by not renewing their contracts, demoting them, freezing their chances of promotion, reproving them about the offending publication, or by a chilly silence about the controversial work.

## The impact of politics on research

Politics influences every aspect of research on fertility control, from the frames of reference within which research is conceived and evaluated to the ways in which findings are presented to, or concealed from, the public.

### Frames of reference

Politics blends into the frames of reference used in research on fertility control. These are criteria defining what is valuable or worthless in research and what constitutes objectivity and bias in the design and interpretation of studies.

For example, a review of a book that challenges the validity of population control (Kasun, 1988) begins with a characterization of the author: "Jacqueline Kasun, an economist, has been actively supportive of efforts by segments of the American right-to-life movement to ban abortion and to terminate public support for population and family planning programs, both domestically and in US foreign assistance" (Crane, 1989: 368–369). The reviewer appears to attack Kasun not because of the arguments in her book, but because of her connections with the right-to-life movement. Crane also criticizes Kasun for being too closely identified with the views of Julian Simon, a vocal opponent of mainstream views on fertility control.

### The choice of research organizations and scholars

The political setting affects decisions about which organizations and individuals are selected to carry out a given study. The listing of topics in funding announcements gives organizations an immediate clue about the kinds of work the funders wish to see completed. Subjects such as the determinants of success in family planning programs or estimates of demand for family planning in developing countries suggest directions of research tied to fertility control. Funding agencies exercise further control over what is done and who does it through their choice of organizations. Grants to some, such as the Center for Population and Family Health at Columbia University, are more likely to generate findings supporting fertility control programs than grants to others, such as the Heritage Foundation.

Politics likewise affects the choice of particular scholars to work on a research project. One screening comes when staff are hired into an organization. By means of employee self-selection and through recruitment decisions, some scholars are brought in and others are not. Radical critics of population control have about the same chance of being hired by Population Action International as advocates of abortion have of being hired to do population research at the US Catholic Conference.

The political setting further influences nominations to prominent research bodies, such as the US Commission on Population Growth and the American Future and the committees on population of the National Research Council. Bachrach and Bergman observed about the US Commission: "While the family planning position was well represented by commissioners from organizations which have supported it—the Population Council, Planned Parenthood/World Population, the Ford Foundation, and the Rockefeller Foundation—neither leading critics of the family planning strategy, nor ecological critics were appointed to or represented on the commission. . ." (1973: 62).

Simon (1986) raised the question of which authors were chosen and why for the National Research Council's 1986 volume on population growth and economic development. During a personal conversation with Simon, the staff director of the NRC reported that authors were chosen as people "who were not known to have a strongly fixed position . . . at one or another end of the continuum" in order to "avoid the group becoming a battleground" (1986: 575). As a result, neither strong critics of population control, such as Simon, nor advocates of stringent control, such as Paul Ehrlich, were invited to be members. The ends of the ideological distribution were cut off by the way in which the drafting group was put together.

### Research design

Research on fertility control requires making decisions about the topic of research; the methodology of data collection; the sample of people or other units; the wording of questionnaires, interview schedules, guides to observation, or other written instruments; how to deal with problems in data collection; and similar matters. Whatever their design, all studies have room for discretionary choices that can be affected by politics.

Among the influences at work in choosing a research topic are the investigator's own sense of priorities about what should be done; the availability of funds; direct or indirect pressures from the researcher's institution; and the necessary contacts and supporting facilities for conducting research in a particular country. Politics can come into play when the researcher makes a personal decision about a topic; when organizations and their staff accept funds from donors to pursue certain studies; when employers signal that some types of inquiry, such as research that is of benefit to family planning, will lead to promotions and other types will not; and when donors and government agencies in developing countries open their doors to investigators. Funding agencies, such as USAID, UNFPA, and the World Bank, strongly influence the choice of topics by making grants available for certain areas of research.

In the 1960s, priority topics were the demand for family planning (Berelson, 1966) and the harmful effects of population growth on economic development (Coale and Hoover, 1958; see also the discussion in Hodgson, 1988). There was then little concern over the impact of economic development on client interest in the practice of family planning; the politics of implementing family planning programs; the strategies and tactics of international donor agencies; and the links between culture and client interest in family planning. In the 1990s, population research covers a broader range of topics, but donor funding is still slanted toward studies with benefits for population limitation.

The politics of research design are evident in the KAP type of survey developed in the 1960s. Stycos was frank about their agenda: "The most important function of such surveys is similar to any market research project: to demonstrate the existence of a demand for goods and services, in this case birth control. . ." (1964: 368). Berelson claimed that KAP studies not only could but did give political legitimacy to family planning: "[T]he KAP survey done in Turkey in 1963 was given wide attention and contributed to bringing about the recent change in national population policy" (1966: 665).

One specific bias in KAP surveys was the notion that reproductive decisions are primarily made by women. That idea has been challenged on the grounds that choices about fertility involve husbands and wives and, in some regions, other family members as well (Godwin, 1973; Okediji, 1974; Warwick, 1983). Two other drawbacks were the assumptions that respondents had thought enough about fertility to have a defined opinion about family size, and that, if they had such an opinion, they would tell it to a stranger who arrived unannounced and stayed a short time.

But the Achilles' heel of KAP methodology was the use of questions that led respondents toward the approval of family planning. A study in rural Thailand (Hawley and Prachuabmoh, 1966: 537–538) included the following items:

> (1) If you knew of a simple, harmless method of keeping from getting pregnant too often (more than you want), would you approve or disapprove of its use?
>
> . . .
>
> (3) Are you interested in learning about how to keep from getting pregnant too often or having too many children (more than you want)?

In a stinging criticism of questionnaire wording in KAP surveys, Anthony Marino remarked:

> Given the use of phrases like "too often", "too many", or "more than you want" as explicit cues to the respondent, we may well marvel not at the fact that a majority approve of "this kind of thing," but that anyone could possibly disapprove. (1971: 47)

The political mission behind KAP surveys led to studies that fell below the technical standards of comparable surveys being carried out simultaneously on other topics. In the 1960s and early 1970s there was some criticism of this research (Hauser, 1967; Marino, 1971; Caldwell and Gaisie, 1971), but it did not lead to marked improvements in design (Warwick, 1983).

Some apologists tried to dispel doubts about the validity of the data. Berelson wrote:

> It is easy to make technical criticisms of such studies, but it is not easy to do better. . . . [I]n my opinion, the sampling has been fairly good, the formulation of questions acceptable or better, the interviewing at least adequate, and the analysis, if anything, overdone (again from the standpoint of administrative requirements). (1966: 657)

Freedman remarked:

> I would accept the responses on such surveys as valid initially until they are tested by a really effective, persistent, all-out service and information effort. Devious psychological explanations of why the respondents did not mean what they said may be too easy a rationalization for a feeble or insufficiently thorough [program] effort. (1966: 815)

Caldwell and Gaisie were more critical:

> KAP surveys in Africa, and doubtless elsewhere, are beset by problems of communication, definition, and clear thinking. The subject is delicate; probing often does not go far enough, especially in such areas as the limitation of sexual relations between spouses. There is often a reluctance to discuss the matters being investigated—a reluctance which can exist between organizer and interviewers as well as between interviewers and respondents. . . . (1971: 57–58)

Caldwell and Gaisie also saw difficulties with the sampling frames used to select respondents, memory lapses, problems in obtaining accurate statements about age, and errors caused by questions that respondents did not understand.

The introduction of the World Fertility Surveys in the 1970s raised the quality of methodology in survey research on fertility in the developing countries (Cleland and Hobcraft, 1985). Sponsors such as USAID may have drawn political benefits from the data, but they supported higher professional standards than had been seen in most KAP surveys.

The politics of population research also left its mark on the *World Development Report 1984* (World Bank, 1984), one of the leading publications on fertility control during the 1980s. Its topics included the consequences of rapid population growth, conditions that lead to reduced fertility, experience with and obstacles to family planning programs, and government population policies. The report concluded that rapid population growth poses grave dangers to economic development, and that public policy should be used to reduce fertility.

The World Bank is one of the most staunch advocates of fertility control. According to Fred Sai and Lauren Chester, "[T]he primary objective of the Bank's work in population is to help slow population growth by reducing fertility and mortality, and thereby permit faster improvements in productivity, GNP growth, and maternal and child health than would be possible if rapid population growth were to continue" (1990: 180).

Reviews suggest that the Bank's suppositions about fertility control affected the research design of the *World Development Report.* Ronald Lee (1985) cites three biases in the data:

(1) The report ignored cross-national studies showing the relationship between population growth or density and rates of income per capita or income distribution. "These studies typically find no relation at all between growth rates of per capita income and population, and sometimes find a positive association of population density and growth rates of per capita income; for income distribution the results are quite mixed" (p. 128).

(2) The analysis dismissed the possibility, advanced by Ester Boserup (1966 and 1981) and Julian Simon (1981), that population size, population growth, and population density contribute to technological progress.

(3) Numerous assertions, such as a statement that rapid growth in the labor force increases unemployment, are based on a priori arguments or no empirical evidence at all.

Lee claims that the report is a cogent position paper for the World Bank's views on population policy. But its value as a scholarly document is limited by its presentation of "assertions for which I am unaware of empirical evidence" (pp. 129–130). Harvey Leibenstein (1985: 136) likewise faults the report for relying too heavily on hypothetical rather than empirical evidence. Richard Easterlin concludes that "this Report is essentially a brief for the World Bank's official position—vigorous population policies now, throughout the developing world" (1985: 115).

The politics of research can force investigators to abandon a viable research design once a project has started. One element in the Project on Cultural Values and Population Policies, a multicountry initiative sponsored by UNFPA, was analysis of the values and operating strategies of the leading international donor agencies in population. Between 1973 and 1977 project researchers carried out interviews with officials in USAID, the Ford Foundation, the Population Council, the Pathfinder Fund, and other donors. The staff prepared draft reports on three agencies and sent them to UNFPA. Worried that this research would harm UNFPA's relationships with the agencies studied, its executive director ruled that no additional UNFPA funds could be spent on studying donors other than the UNFPA itself (Warwick, 1987).

### Interpreting data

An advocacy bias often shows itself in the interpretation of research data. In its chapter on the consequences of rapid population growth, the *World Development Report 1984* states:

> [T]he evidence discussed above points overwhelmingly to the conclusion that population growth at the rapid rates common in most of the developing world slows development. At the family level . . . high fertility can reduce the amount of time and money devoted to each child's development. It makes it harder to tackle poverty, because poor people tend to have large families, and because they benefit less from government spending on the programs they use most—health and education, for example—when public services cannot keep pace with population growth. At the societal level. . . it weakens macroeconomic performance by making it more difficult to finance the investments in education and infrastructure that ensure sustained economic growth. (1984: 105)

Two scholars argue that these conclusions do not follow from any evidence presented in the report. Leibenstein (1985) notes that the impact of population on economic welfare is unknown. The report, in his view, tries to fill the gap in empirical evidence with simulations and projections, but the result is unconvincing. Lee (1985) challenges the evidence offered to support the main conclusion about the harmful effects of population growth and takes issue with statements about several specific consequences. Like Leibenstein, he holds that "sometimes one cannot distinguish statements that are hypotheses or speculation from those that reflect the conclusions of an empirical literature" (p. 129).

### Publication

Politics bears on whether research findings are published, where, when, and how. Two personal examples illustrate this point. In one I had prepared a draft paper on an Asian country's family planning program. Because that country's government had sponsored the research and provided much of the information on which it was based, I sent it to the program's head for comments. He expressed concern that, if published, sections of the paper could do political harm to him and his organization. At about the same time, colleagues from my own research institute, which had various government-sponsored projects in the same country, prepared a review of the paper. They sent copies of their comments to me and complained about the paper to the director of the institute. Representatives of an international donor agency, who had not read the paper, also warned the journal in which it eventually appeared about their concern with what it might say. Anonymous reviews commissioned by the journal proved to be much more favorable than those from my colleagues, and the paper was published without delay. It apparently caused no damage, and, in a telex to me, another senior program official praised the essay for its clear analysis.

In the Project on Cultural Values and Population Policies, which I managed, UNFPA decided that the research reports were too controversial to be published by that agency. The official in charge gave this explanation:

> This review confirmed . . . that the proposed publication would not be likely to be useful for guidance in population policy making in United Nations agencies or governments. Furthermore, the text in a number of places stated positions and conclusions with which we could not associate ourselves. We, therefore, are not prepared to publish the proposed brochure, either as an official publication of the Fund, or as a document otherwise officially associated with the Fund. (Warwick, 1987: 168)

The main findings were later published in a book (Warwick, 1982).

### Press releases and media presentations

In some projects, dissemination of research results takes place through press conferences and other media events. There, interested parties may overstate findings to capture public attention.

Simon (1986) challenged the publicity given the 1986 study by the US National Research Council on population growth and economic development. The following headline in the press release, he claimed, made population growth seem of greater moment than it was shown to be in the report: "Slower Population Growth Generally Benefits Developing Nations' Economies: Is One of Several Key Factors Cited."

> The word "key" in the headline of the press release was placed there by the staff together with the Office of Public Affairs, without consultation with D. Gale Johnson, Samuel Preston, or Ronald Lee of the Working Group, though Preston characterizes that term as "not a fair adjective" . . . and Johnson seemed surprised when I told him that "key" appeared in the headline. This wording may have been wholly a response to the natural desire to make the matter newsworthy. But such a change in wording might also seem responsive to such elements of the situation as the comment by Congressman Sander Levin (a principal in the population movement) at the rehearsal the night before

> the NAS-NRC presentation that the presentation was "not dramatic enough." (Simon, 1986: 575)

Politics most strongly affects media statements through the simplification of data for public consumption. In the *World Development Report 1984* and in press releases from organizations such as Zero Population Growth and the Population Institute (Fornos, 1989), the simplification is in one direction: the need for slowing population growth.

## The political uses of population research

Just as politics influences research, research can serve politics. Studies of fertility control have five main political uses.

The first is publicizing a particular cause or point of view. Advocates and opponents of induced abortion regularly cite opinion surveys to call attention to the need for more abortion services or to the harm being done by existing services. Recognizing the potential of the media for shaping opinion, the hierarchy of the Roman Catholic Church in the United States has recently decided to hire public relations consultants to help disseminate their position on abortion. To dramatize the harms done by population growth, organizations such as the Population Institute and Population Action International present dozens of reports, newsletters, and testimony to Congress (see Fornos, 1993) summarizing research and other information.

A second political use of research is legitimation or delegitimation. The aim is to show that a certain course of action, such as fertility control, is justified or unjustified. This strategy begins with the very language in which the issues are stated. Those who favor fertility control prefer the phrase family planning, which has overtones of rationality, self-direction, and human welfare, rather than birth control, which implies limits on free choice. Linguistic euphemisms are particularly common in debates about abortion. Those in favor call themselves pro-choice rather than pro-abortion; those opposed, pro-life rather than anti-abortion. Research results on the vacuum aspirator speak of menstrual regulation, which seems benign and medically respectable, rather than early abortion, which is more controversial. In research on fertility control, language is politics.

A common strategy in seeking and challenging legitimacy is to base arguments on science. The results of KAP surveys and their successors were and are used to justify initiatives on fertility control. As Notestein observed (1982: 684), research not only provides findings that can be held up as truth, but, by undertaking action in a delicate field, it legitimates further interventions. Works critical of population programs, such as the studies by Mamdani (1972) on India or by Hartmann (1987) on several countries, use science to undermine efforts at fertility control.

The third use of population research is in implementation. A perfect example of how research findings were utilized to carry out a program comes from action-oriented field research in Taichung, Taiwan. Cernada and Sun (1974) show how their studies served as a base for developing a national family planning program. A careful sample survey as well as studies of field experiments suggested ways in which the program could reach more clients. The research likewise served as an early warning system to identify failures and problems. Close observation of how programs are managed and received across different cultural regions of a country can also provide valuable data for implementation (Warwick, 1988). Some kinds of implementation studies, such as those showing low client demand or widespread political and religious opposition to family planning, can lead managers to slow down the program's speed until conditions improve.

Fourth, research on fertility control can be used for neutralization. This takes place when data are presented to demonstrate that fertility control is safe, normal, and socially acceptable. One method is to display statistics indicating that family planning services are widely used. These convey the message that family planning has become common and respectable. Another approach is to cite medical research documenting the safety of such contraceptive methods as the pill or the intrauterine device. Such information helps to ward off the fears of potential users about harm from contraceptives. The suppression of controversial findings about fertility control programs, such as those showing physical harm resulting from contraceptive use, sterilization, and abortion, also contributes to neutralization.

The fifth political use of research is polarization, which occurs when research rouses hostilities or divides people into contending groups. Kleymeyer and Bertrand (1983) give an example of polarization set off by a research project on family planning in a poor Latin American neighborhood. The main charge was that North American social scientists planned to sterilize local children as part of a government-sponsored antimeasles vaccination campaign. The accusation came alive when critics went to the researchers' rooms and found a phial of vaccine marked STERILE. They then paraded through the city with a loudspeaker telling residents not to let their children be sterilized.

## Conclusion

Politics and research feed into each other. Politics affects the frames of reference for thought in the field, the selection of organizations and individuals to carry out studies, the choice of topics, the interpretation of data, publication, and media presentations. Research supports politics by publicizing issues, improving or holding back implementation, legitimating or delegitimating certain lines of action, and neutralizing or polarizing debate. Politics and research are thus inseparable in studies of fertility control.

## References

Aird, John S. 1990. *Slaughter of the Innocents: Coercive Birth Control in China*. Washington, DC: AEI Press.

Bachrach, Peter, and Elihu Bergman. 1973. *Power and Choice: The Formulation of American Population Policy*. Lexington, MA: Lexington Books.

Berelson, Bernard. 1964. "Turkey: National survey on population," *Studies in Family Planning* 1, 5: 1–5.

———. 1966. "KAP studies on fertility," in *Family Planning and Population Programs*, ed. Bernard Berelson. Chicago: University of Chicago Press, pp. 655–668.

Boserup, Ester. 1966. *The Conditions of Agricultural Growth*. London: Allen and Unwin.

———. 1981. *Population and Technological Change*. Chicago: University of Chicago Press.

Caldwell, John C., and S. K. Gaisie. 1971. "Methods of population and family planning research: Problems in their application in Africa," *Rural Africana* 14, Spring: 53–60.

Cernada, George, and T. H. Sun. 1974. "Knowledge into action: The use of research in Taiwan's family planning program," East-West Center, Paper No. 10. Honolulu, HI: East-West Communication Institute.

Cleland, John, and John Hobcraft (eds.). 1985. *Reproductive Change in the Developing Countries: Insights from the World Fertility Survey*. New York: Oxford University Press.

Coale, Ansley J., and Edgar M. Hoover. 1958. *Population Growth and Economic Development in Low-Income Countries*. Princeton: Princeton University Press.

Crane, Barbara B. 1989. Review of Jacqueline Kasun, *The War Against Population: The Economics and Ideology of Population Control, Population and Development Review* 15, 2: 368–370.

Demeny, Paul. 1988. "Social science and population policy," *Population and Development Review* 14, 3: 451–479.

Easterlin, Richard A. 1985. Review of The World Bank, *World Development Report 1984, Population and Development Review* 11, 1: 113–119.

Fornos, Werner. 1989. "Population bomb keeps ticking," *The New York Times*, 14 October.

———. 1993. "Testimony of Werner Fornos, President, The Population Institute," in *Hearings before the Subcommittee on Foreign Operations, Export Financing, and Related Programs, Committee on Appropriations, House of Representatives, One Hundred Third Congress, First Session, Part 3*. Washington, DC: US Government Printing Office, pp. 162–177.

Freedman, Ronald. 1966. "Family planning programs today: Major themes of the conference," in *Family Planning and Population Programs*, ed. Bernard Berelson. Chicago: University of Chicago Press, pp. 811–825.

———. 1987. "The contribution of social science research to population policy and family planning program effectiveness," *Studies in Family Planning* 18, 2: 57–82.

Godwin, R. Kenneth. 1973. "Methodology and policy," in *Population and Politics*, ed. Richard L. Clinton. Lexington, MA: D.C. Heath, pp. 131–143.

Hartmann, Betsy. 1987. *Reproductive Rights and Wrongs: The Global Politics of Population Control and Contraceptive Choice*. New York: Harper and Row.

Hauser, Philip M. 1967. "Family planning and population programs: A book review article," *Demography* 4: 402–405.

Hawley, Amos H., and Visid Prachuabmoh. 1966. "Family growth and family planning in a rural district of Thailand," in *Family Planning and Population Programs*, ed. Bernard Berelson. Chicago: University of Chicago Press, pp. 523–542.

Hodgson, Dennis. 1988. "Orthodoxy and revisionism in American demography," *Population and Development Review* 14, 4: 541–569.

Kasun, Jacqueline. 1988. *The War Against Population: The Economics and Ideology of Population Control*. San Francisco: Ignatius Press.

Kleymeyer, Charles D., and William E. Bertrand. 1983. "Misapplied cross-cultural research: A case study of an ill-fated family planning research project," in *Social Research in Devel-*

*oping Countries: Surveys and Censuses in the Third World,* eds. Martin Bulmer and Donald P. Warwick. Chichester, United Kingdom: John Wiley and Sons, pp. 365–378.

Lee, Ronald. 1985. Review of The World Bank, *World Development Report 1984, Population and Development Review* 11, 1: 127–130.

Leibenstein, Harvey. 1985. Comment on The World Bank, *World Development Report 1984, Population and Development Review* 11, 1: 135–137.

Mamdani, Mahmood. 1972. *The Myth of Population Control: Family, Caste and Class in an Indian Village.* New York: Monthly Review Press.

Marino, Anthony. 1971. "KAP surveys and the politics of family planning," *Concerned Demography* 3, 1: 36–75.

McCoy, Terry L. (ed). 1974. *The Dynamics of Population Policy in Latin America.* Cambridge, MA: Ballinger Publishing.

Notestein, Frank W. 1971. "Reminiscences: The role of foundations, the Population Association of America, Princeton University and the United Nations in fostering American interest in population problems," *The Milbank Memorial Fund Quarterly* 49, 4, Part 2: 67–84.

———. 1982. "Demography in the United States: A partial account of the development of the field," *Population and Development Review* 8, 4: 651–687.

Okediji, Francis Olu. 1974. *Changes in Individual Reproductive Behavior and Cultural Values.* Liège: International Union for the Scientific Study of Population.

Pathfinder Fund. 1990. "Population organizations work together," *Pathways* 5, 1: 1–2.

Piotrow, Phyllis Tilson. 1973. *World Population Crisis: The United States Response.* New York: Praeger.

Sadik, Nafis. 1990. "The role of the United Nations—From conflict to consensus," in *Population Policy: Contemporary Issues,* ed. Godfrey Roberts. New York: Praeger, pp. 193–206.

Sai, Fred T., and Lauren A. Chester. 1990. "The role of the World Bank in shaping Third World population policy," in *Population Policy: Contemporary Issues,* ed. Godfrey Roberts. New York: Praeger, pp. 179–191.

Simon, Julian L. 1981. *The Ultimate Resource.* Princeton: Princeton University Press.

———. 1986. Review of National Research Council, *Population Growth and Economic Development: Policy Questions, Population and Development Review* 12, 3: 569–577.

———. 1990. "The population establishment, corruption, and reform," in *Population Policy: Contemporary Issues,* ed. Godfrey Roberts. New York: Praeger, pp. 39–58.

Stycos, Joseph Mayone. 1964. "Survey research and population control in Latin America," *Public Opinion Quarterly* 28, 3: 367–372.

United Nations Population Fund (UNFPA). 1992. *Global Population Assistance Report 1982–1990.* New York: UNFPA.

Warwick, Donald. P. 1982. *Bitter Pills: Population Policies and Their Implementation in Eight Developing Countries.* New York: Cambridge University Press.

———. 1983. "The KAP survey: Dictates of mission versus demands of science," in *Social Research in Developing Countries: Surveys and Censuses in the Third World,* eds. Martin Bulmer and Donald P. Warwick. Chichester, United Kingdom: John Wiley and Sons, pp. 349–363.

———. 1987. "The politics of population research with a UN sponsor," in *The Research Relationship: Practice and Politics in Social Policy Research,* ed. G. C. Wenger. London: Allen and Unwin, pp. 167–184.

———. 1988. "Culture and the management of family planning programs," *Studies in Family Planning* 19, 1: 1–18.

World Bank. 1984. *World Development Report 1984.* New York: Oxford University Press.

# TRANSNATIONAL ACTORS AND FAMILY PLANNING POLICY

# Population Policy and Feminist Political Action in Three Developing Countries

Ruth Dixon-Mueller
Adrienne Germain

The years since 1975 have witnessed the rise of a genuinely international women's movement (Tinker and Jaquette, 1987; Shreir, 1988; Wieringa, 1988). Stimulated in part by the resurgence of feminist movements in Europe and North America in the mid-1960s (Basnett, 1986; Duchen, 1986), and by the efforts of the Commission on the Status of Women at the United Nations, the UN designation of 1975 as International Women's Year not only raised the consciousness of women around the world, but also served as a dynamic organizing tool. The decade between the 1975 women's conference in Mexico City and the 1985 conference in Nairobi also witnessed a remarkable growth in the awareness, the militancy, and the organizational networks among women in developing countries (Mies, 1986). Individual women and organized women's groups in developing countries—as in the industrialized world—are critical of prevailing social and economic policies and practices (see, for example, Sen and Grown, 1987), and they are demanding to be heard in many of the public debates and political decisions from which they have largely been excluded. Women's groups are also emerging as vitally important to the success and the legitimacy of the public policies that affect them.

Policy decisions relating to childbearing and the delivery of family planning and health services have the potential to affect the lives of women of all social classes in fundamental ways. Whether policies have a positive or negative impact may depend on the integration of women from diverse backgrounds into the decisionmaking process. Whether or not women are recognized—and recognize themselves—as a primary constituency in population and family planning decisions is, therefore, a key public policy issue.

The impact of women's political action on population policies and family planning programs depends, in part, on how open the decisionmaking process is to participation by women's rights advocates and on the extent to which governmental and nongovernmental agencies consider women an important constituency. The effect of women's activism depends also on whether or not women's groups act collectively to exert pressure on key agencies.[1] Will women from different backgrounds present themselves as separate interest groups or as a collective movement to be reckoned with in the national political arena? The activism of women's organizations in each country, the political autonomy of such organizations, the emergence of a shared ideology and agenda in the form of a women's movement, and the position of reproductive rights and women's health as political issues all contribute to women's capacity to effect change. Whether pressure tactics can be used effectively will depend on the nature of the relationship between the women's movement and bureaucratic and legislative institutions, political parties, religious institutions, and medical and family planning organizations.

In the light of these considerations, this chapter examines the impact of women's political action on population, family planning, and health policies during the 1980s in Brazil, Nigeria, and the Philippines. These three countries have been selected because they represent similar processes of democratization and economic crisis but different regions, population policies, demographic processes, and levels of women's political mobilization.

All three countries have undergone a process of democratization since the mid-1980s that permits us to examine women's political action under evolving conditions of popular participation. Brazil and the Philippines have emerged from long periods of repressive authoritarian rule, and Nigeria's military government has lifted its ban on partisan politics as part of a planned transition to civilian rule. These more open political environments have brought population policies, along with other critical issues, to the forefront of public debate as the new governments seek to legitimize their policies in the eyes of key political constituencies. In turn, women's groups and other interest groups and institutional actors have been seeking to define, influence, or challenge the public discourse on these issues. The governments of all three countries are experiencing international debt crises, however, resulting in the imposition of economic austerity measures that have created considerable unrest. Economic difficulties and resulting political tensions impede each government's ability to improve the quality and quantity of family planning and health services, even where they are committed to doing so.

The three countries also represent contrasts that are relevant to this analysis. Brazil's income per capita of about US$2,300 in the late 1980s, which is highly concentrated in the hands of the wealthy, is almost four times higher than that of the Philippines and eight times higher than that of Nigeria. Brazil's population growth has dropped to about 2.0 percent annually, compared with a fairly stable 2.6 percent in the Philippines and almost 3.0 percent in Nigeria. Brazil does not have an official policy to reduce population growth, whereas the Philippines has maintained such a policy since 1971, and Nigeria announced its first policy to encourage lower fertility in the context of national development in 1989.

The case studies we present also reveal different configurations of feminist political activism. In Brazil, an active and explicitly feminist movement, capitalizing on its political support for the democratic coalition that defeated the military regime, has moved creatively to influence governmental policies on integrated women's health and family planning programs, with some effect. Nigerian women, organized into multiple associations but lacking a coherent women's movement, have played a predominantly reactive role in their response to the new national population policy, with little effect. In the Philippines, the nascent women's movement arising out of the return to democracy has been creatively reactive in its efforts to protect women's access to family planning services (and, indirectly, the national population program) from attacks by conservative forces, with some effect. The three country scenarios presented here are, therefore, unique but not unrelated.

## Brazil: Feminist critique and feminist influence

The women's movement in Brazil is dynamic, politically active, generally leftist, and explicitly feminist in its focus on gender as well as class relations: Women's sexual and reproductive rights are integral to its policy agenda.[2] The movement has evolved from a complex set of social, economic, and political conditions. These include a critical awareness among feminist intellectuals of the dynamics of unequal distribution of wealth and other resources in Brazilian society, a prolonged struggle against two decades of authoritarian rule (which ended in 1985), and the intense involvement of some elite women—in political exile during the 1970s—with feminist movements in Europe, especially in France (Schmink, 1981; Corrêa, 1989; Sarti, 1989).

As in most countries, the celebration in Brazil of International Women's Year in 1975 both stimulated and legitimated research and action on a wide range of women's issues such as working conditions, legal rights, child care, and sexual violence. In 1978, with the democratic reopening under the military regime of some aspects of the political process, and with the reintroduction of political parties, women's groups flourished in many forms: feminist collectives, neighborhood residents' associations, mothers' clubs affiliated with the progressive wing of the Catholic Church, political groups, and women's associations within the professions and trade unions. Despite their diversity, women's groups were mobilizing as a collective movement to influence state policies (Instituto de Ação Cultural, no date; Sarti, 1989).

As opportunities for direct political participation opened, women who remained apart from the political process in order to protect the autonomy of their movement were challenged by those urging active involvement in electoral party politics (Blay, 1985; Alvarez, 1989). Feminists worked enthusiastically with opposition candidates in several state elections of 1982 to incorporate women's rights into their party platforms. Ultimately, the government of Tancredo Neves, elected in 1985, owed a considerable amount of its popular support to organized women's constituencies (Alvarez, 1989).

### Reproductive health and the feminist critique

Brazilian feminists have developed what they call a "critical understanding of the conditions of human reproduction in Brazilian society" (Barroso, 1990; Corrêa, 1989; Sarti, 1989). Joined by some likeminded physicians (see Pinotti and Faúndes, 1989), the feminists presented a critique of prevailing reproductive health policies and practices in Brazilian society that includes the following points:

First, the quality of primary health care for women in Brazil is generally poor, especially for rural women and those in the urban *favelas* (slums), who cannot afford private care. Reflecting extreme inequalities in the distribution of incomes and health services, rates of maternal and infant mortality and morbidity remain among the highest in Latin America despite Brazil's relatively high income per capita.

Second, in the past the major private family planning organizations have appeared more interested in recruiting contraceptive acceptors than in protecting women's physical and emotional health and their right to informed choice.

Third, contraceptives are not available in most government health services, despite the government's statement at the 1974 World Population Conference in Bucharest that "the State [should] provide the information and the means [of birth control] that may be required by families of limited income" (Merrick and Graham, 1979: 281). Most Brazilian women have a limited choice of birth control methods and service providers, as attested by their heavy reliance on private physicians and pharmacies and on three methods only: the pill, sterilization, and clandestine abortion.

Fourth, abortion is illegal in Brazil except to save a woman's life or in the cases of rape. Illegal services provided in expensive private clinics are safe, but poor women often resort to dangerous self-induced or backstreet procedures. In 1980, more than 200,000 women were hospitalized for treatment of incomplete or septic abortions in health clinics run by the medical branch of the national social security system, which serves about 80 percent of Brazil's population. An estimated three to four million women resort to illegal induced abortions each year (Pinotti and Faúndes, 1989).

Fifth, feminists believe that levels of female sterilization (more than 27 percent of currently married women nationally) are excessive, especially in low-income groups, reaching more than 40 percent in the North and Central-West regions (United Nations, 1988a). Legal restrictions on sterilization have encouraged the practice of cesarean sections at childbirth (almost one-third of all deliveries), because sterilization can be performed inconspicuously at this time and because women with a history of cesareans can claim medical grounds for sterilization (Rutenberg and Ferraz, 1988).

Finally, feminists have charged that "both natalist and antinatalist politics have utilized sexuality, the body of woman, as a social patrimony, denying her rights and her individuality" (Alvarez, 1989: 216). Opposed to the pronatalist stance of Brazilian nationalists and the Catholic Church hierarchy, as well as to the antinatalist policies of the family planning organizations and their allies in government circles, feminists insist on the primacy of a woman's right to control her own sexuality and reproduction according to her personal needs and experiences.

### Feminist influence on state policy

At the time of the World Population Conference in 1974, the Brazilian government was encouraging population growth as a means of settling the country's vast frontier regions and fueling its economic advance and international prestige. As the economic miracle of the late 1960s and early 1970s metamorphosed into the widespread unemployment and debt crisis of the 1980s, however, concerns about the impact of population growth loomed larger on the political agenda. A federal plan to curb population growth was drafted by sectors of the military in consultation with internationally funded private family planning organizations that were pressing for adoption of population-control measures (Alvarez, 1989). In June of 1983, during the waning months of the military regime, the chief of the armed forces spoke out publicly on the "utmost importance" of a national family planning program. Many religious, intellectual, and government leaders were opposed to such a program, however, for a variety of reasons, and the plan was not approved (United Nations, 1988a).

When 21 years of authoritarian rule ended with the election of a civilian federal government in January 1985, the time seemed ripe for women activists to play a more direct role in state and federal policy formation. As Carmen Barroso notes (1990: 10–11), "For a great part of the feminist movement it was an exciting challenge to try to affect public policies they had been criticizing from a safe distance." To be successful, however, feminists had to learn how to translate their radical critiques into feasible programs, how to gain political leverage, and how to manage the complex federal, state, and municipal bureaucracies with which they would work (see Blay, 1985).

Appointed by the new government to several high-level national councils, Brazilian feminists were mandated to make broad-ranging recommendations on reproductive health policies and programs and to promote their implementation. Feminists made significant contributions to the national Integrated Women's Health Program adopted by the Ministry of Health in 1983 as an alternative to the military's population-control plan; to the Policy on Women's Health and Family Planning adopted by the Ministry of Social Security in 1986, which mandated for the first time that the state should provide a full range of birth control methods through its social security health services under conditions of fully informed consent; to the National Council for Women's Rights appointed by President José Sarney in August 1985; and to the Committee on Reproductive Rights attached to the Ministry of Health.

The National Council for Women's Rights actively promoted women's representation in the Constituent Assembly, which was elected in November 1986 to write a new Brazilian constitution. Several of the 26 women elected to the 559-member body used the opportunity to advance an explicitly feminist agenda on reproductive rights. A significant number of members of the assembly, however, activated by evangelical churches and the North American "right-to-life" movement and supported by the conservative wing of the Catholic Church hierarchy, sought through popular petition and intensive lobbying to introduce a fetal protection clause in the new constitution (Barroso, 1990). Feminists responded with demonstrations and a popular petition for legalized abortion on demand. Ultimately, feminists and their supporters blocked the adoption of a constitutional ban on abortion by arguing that issues that had not been publicly debated should not be included in the constitution. The abortion fight symbolized the collapse of what had once been a strategic alliance of the women's movement, the Catholic Church, and the political left around grassroots struggles for social justice under the authoritarian regime (Sarti, 1989).

The dialogue between feminists and the state and political parties has resulted in the partial institutionalization of a feminist perspective on women's health and rights at the federal level. In this and other respects, the Brazilian women's movement has exemplified what can be attempted and accomplished for feminists engaged in similar campaigns in countries such as Colombia, Mexico, Nicaragua, Peru, and Venezuela.[3] But a number of obstacles remain. Implementation of the women's health program has proved difficult, because of the economic crisis, the decentralization of health planning to the state level, and a lack of political will. The continued involvement of private family planning agencies, combined with opposition on the part of the Catholic hierarchy to the provision of artificial methods of birth control in state health facilities, has reduced the pressure on the state to provide public services. In addition, political leadership at the federal level has become more conservative, and feminist influence is fading. In response, activists are concentrating on influencing policies and programs in key states and municipalities such as São Paulo.

Brazilian feminists nevertheless remain committed to establishing, across the broad spectrum of society, the principle of full and democratic citizenship that is free of gender bias in all its forms. The achievement of reproductive and sexual rights—including freedom from violence against women—is seen as central to this goal. Three principal directions of feminist influence have been proposed in this context (Barroso, 1989): (1) a radical redefinition of the structure and quality of women's health services; (2) a continued struggle to institutionalize women's power and voice in policy making; and (3) the decriminalization of abortion in recognition of women's capacity to decide consciously and responsibly about their bodies, their sexuality, and their lives.

## Nigeria: In search of a reproductive health agenda

Nigerian women throughout their history have engaged in a variety of actions such as mass protests and strikes to protect their interests—particularly their economic interests—against encroachments by the state and other ruling elites (Mba, 1982; Okonjo, 1983). They are used to having their voices heard. Famous for their community leadership and dynamic market women's association, Nigerian women of all social sectors mobilize in multiple and overlapping networks around cultural, social, economic, philanthropic, occupational, and religious interests (Mba, 1982; Enabulele, 1985; Kisekka, 1989b). Autonomous women's groups represent a broad range of socioeconomic and educational levels ranging from the nonliterate (for example, farm women's cooperatives and village-based mutual-aid societies) to the educated and affluent (for example, such internationally affiliated urban groups as the Nigerian Association of University Women, Soroptomists, and Zonta). Although most associations are independent of men's influence, exceptions include the clubs of wives of professional men, especially those in the armed forces, and the women's wings of political parties.

The largest organization of Nigerian women, the National Council of Women's Societies (NCWS), was launched in 1959 with the merging of two national societies in order to promote women's participation and welfare in social and community affairs. Its leadership draws significantly from wives of important bureaucrats, businessmen, and politicians. From its earliest days, NCWS proclaimed itself (and was regarded by the government) as the organization representing all Nigerian women (Mba, 1982). During the years from 1966 to 1979, when the previous military regime banned all political party activity, only the NCWS continued to function as a national women's interest group (Mba, 1982). The council has generally played a centrist role vis-à-vis the state in complementing and supporting government initiatives, urging the government to name women to various councils and bodies.

As in Brazil, the celebration of International Women's Year in 1975 activated new constituencies of Nigerian women in universities, professions, and businesses around issues of legal rights and women's roles in social and economic development (Ogunsheye et al., 1988). Some new groups that formed in the 1980s include members who define themselves as feminists, such as Women in Nigeria, a university-based radical feminist and leftist organization, founded in 1982, and the Women's Health Research Network of Nigeria, begun in 1988 (Kisekka, 1989a, 1989b, 1989c). In general, however, the feminist voice has been muted; and, despite the ubiquity of women's organizations in all segments of society and their activism on various local issues, there is no clearly identifiable, autonomous women's movement in Nigeria with a collective national strategy for change.

### The new population policy and reactions of women's groups

Nigeria did not have an official policy on slowing population growth prior to the government's formal adoption of a National Policy on Population for Development, Unity, Progress, and Self-Reliance in 1989. The Nigerian delegation at Bucharest in 1974, like its Brazilian counterpart, expressed confidence that the country could absorb its population increase—which may by then have reached 3.0 percent annually—without difficulty (Orubuloye, 1983). By the mid-1980s, however, the Nigerian economy was worsening following the oil-induced boom years of the 1970s, income per capita and agricultural output were dropping (Federal Republic of Nigeria, 1988), and economic austerity measures were imposed. In 1985, the military government introduced the draft of a population plan devised by the federal Ministry of Health with the participation of other ministries and several international donor agencies and family planning organizations (United Nations, 1988b). For three years, drafts were reviewed and revised by various professionals, religious leaders, politicians, and other groups as part of a national consensus-building process. Containing unusually explicit goals and objectives, the national policy was officially launched by President Ibrahim Babangida in early 1989.

Claiming they had not been consulted in the consensus-building process, the major national women's organizations were quick to respond. "It is no longer acceptable for planners and policymakers to develop policies and plans targeting specific communities of interest without including those communities through representative community organizations," charged the president of NCWS (*New Nigerian*, 1989). Coming from this long-established organization representing several hundred women's associations throughout Nigeria, the charge of neglect of women's interests was serious. Representatives of NCWS, the Federation of Muslim Women's Associations (which was established in the North in 1985), and other women's organizations reacted strongly—but by no means unanimously—to several provisions of the new policy.

First, representatives of women's groups protested widely in the press that the target of reducing fertility to four children per woman by the year 2000 (from about seven in the mid-1980s) was discriminatory, because men were merely "encouraged to have [the] limited number of wives and [the] optimum number of children they can foster within their resources" (Federal Republic of Nigeria, 1988). The four-child-per-woman limit also sparked the emotionally charged atmosphere of demographic rivalries among religious, ethnic, and regional groups (compare Kokole, and Panandiker and Umashankar, in this volume), because Muslim men and others practicing polygyny were not bound by the four-child rule.

Second, the policy explicitly states that, "In our society, men are considered the head of the family and they take far-reaching decisions including the family size, subsistence and social relations. . . . The patriarchal family system in the country shall be recognized for stability of the home" (Federal Republic of Nigeria, 1988: 19). Feminists were particularly upset by this statement because, by implication, it formalizes men's right to control women's sexuality and childbearing under the guise of maintaining the stability of the family.

Third, as noted earlier, the president of the NCWS protested that women's associations are expected to support and promulgate the policy at the state and local levels but that they were not included in its formulation. Others questioned whether the policy will be "a real blueprint for action or whether it will simply gather dust" because it does not include specific recommendations for implementation (Kisekka, 1989: 6).

Lastly, conservative representatives of Muslim women's groups, many of whom observe traditional Islamic behaviors such as *purdah* (the seclusion of women from public view), opposed the recommendations for educating young people in the schools and the community on population and family life, because they considered such education to be a parental responsibility within the context of religious and moral training. Some also opposed, as inconsistent with traditional Islamic family values, the goals of eliminating arranged marriages of young girls, postponing average marriage age to at least 18 years for females and 25 years for males, and avoiding pregnancies before age 18.

Despite their criticisms, women who have spoken out on the population policy, including NCWS representatives and feminists, have supported certain of its provisions. These include providing universal voluntary access to family planning services, reducing maternal and child mortality, and creating intensive action programs aimed at improving the situation of women and girls in accordance with the provisions of the UN Convention on the Elimination of All Forms of Discrimination Against Women, to which Nigeria is signatory (Federal Republic of Nigeria, 1988).

The proposal to end discrimination against women is consistent with the government's expressed commitment to the advancement of women within the context of national social development (for an alternative view, see Dennis, 1987). Some prominent women have been appointed to state and federal offices, and an institutional base for women's programs has been created through initiatives such as the Primary Health Care Program, the National Social Development Policy, the Commission on Women and Development, and the Better Life for Rural Women Program.[4] The stage is set for specific input on reproductive health programs from national, state, and local women's organizations. Most women's groups are preoccupied with other issues, however, and remain differentiated, if not divided, in their world views by ethnic, religious, regional, and occupational interests, many of which bear on issues of population policy and reproductive health.

### Defining a reproductive health strategy

The concept of reproductive health illuminates the social value placed on women's reproductive capacity and child survival in Nigeria. Yet, in the past, women's reproductive health problems—with the partial exception of prenatal care—have not been given priority in public health services. By default or design, most women have sought care primarily from indigenous healers, pharmacists, traditional birth attendants, and spiritualists, or else have suffered in silence.

According to estimates of the Nigerian Demographic and Health Survey, only 4 percent of married women of reproductive age in 1990 were using modern methods of contraception (Nigeria, Federal Office of Statistics, 1992). Yet, traditional means of child spacing, such as prolonged breastfeeding, postpartum abstinence, and residential separation of married couples, have been eroding. Despite recent mortality declines, maternal and infant death rates remain high (World Health Organization, 1986). Sterility or subfecundity, often resulting from sexually transmitted diseases or from infection caused by unhygienic delivery or abortion, stigmatizes women, because female adult identity and social status depend heavily on bearing children to enrich and empower the male lineage (Women in Nigeria, 1985; Ogunsheye et al., 1988).

The marginalization of reproductive health issues in most women's associations may be due in part to what Mere Kisekka (1989b: 16) calls an ethic of "nobility in suffering," which is inculcated in young girls. The dictates of cultural or religious modesty, combined with a tendency to place women's own health-care needs last, discourage women from speaking out about sexual and reproductive problems. Women's groups that raise sensitive issues, such as the need for contraceptive services for unmarried adolescents or for information to prevent the spread of sexually transmitted diseases, may be criticized as "immoral" or "antifamily" and opposed by other women's organizations. The abortion issue is a good example.

Abortion in Nigeria is a criminal offense for both the woman and the abortion provider except when required to preserve the woman's life (Okagbue, 1990). Maternity wards in many hospitals are crowded with girls and women suffering from incomplete or septic abortions, and many are beyond saving (Ladipo, 1989). Advocates of abortion law reform, such as the Society of Gynecologists and Obstetricians, the Nigerian Medical Association, leaders and members of such professional societies as the Medical Women's Association, the Nurses and Midwives Association, and the Federation of Women Lawyers, as well as feminists, recognize that safe services could significantly reduce prevailing high rates of maternal morbidity and mortality both among married women and among the growing numbers of adolescents in urban and rural areas who are experiencing unwanted pregnancies (Nichols et al., 1986). In 1981, however, the leadership of the National Council of Women's Societies lobbied to prevent a liberal abortion bill from being presented to the House of Representatives (Ilumoka, 1989) on the grounds that parents and the government should promote moral and religious education to prevent pregnancies among unmarried women, and that the government should improve family planning education and services for married women. No further steps have been taken to liberalize the law, despite a widespread perception that clandestine abortion is increasing.[5]

In an attempt to overcome the social stigma and inherent divisiveness of some reproductive health problems and to identify a unifying programmatic theme, a group of social scientists, together with health and legal professionals and community leaders, formed a new organization in 1988 called the Women's Health Research Network of Nigeria. Their purpose is to engage in research and advocacy and to mobilize women's associations, service providers, and policymakers around women's health issues (Kisekka, 1989c).

Two themes are emerging from the network's deliberations that may transcend the diversity of views among women's associations. The first is the prevention of infertility, which afflicts significant numbers of women (and men) in Nigeria as well as in other areas of sub-Saharan Africa (Sherris and Fox, 1983). Infertility is regarded as a tragedy for women who experience it and a source of fear among those who have not yet borne their desired number of children. From a tactical viewpoint, infertility is also a rubric under which to discuss publicly such difficult issues as sexually transmitted diseases, the sexual exploitation of women, septic abortion, women's vulnerability to divorce or replacement by another wife, and harmful reproductive practices. The National Association of Nurses and Midwives, for example, has campaigned against early arranged marriage and sexual intercourse with young girls, female circumcision and infibulation, and the *gishiri* cut (in which traditional midwives cut the birth passage in order to ease a difficult delivery)—practices that can result in excessive bleeding, infection, infertility, vesicovaginal fistula, and even death. The second theme is to ensure women an informed choice among safe, acceptable, and affordable contraceptive methods as a component of basic health care that emphasizes child spacing, safe delivery, and child survival. Consistent with the government's health and population policies, this emphasis requires collaboration among women's associations and health and family planning providers in the public and private sectors to improve both the quantity and quality of care.

The question remains, however, whether a relatively small group of researchers and activists can pull together the threads of a variegated, fragmented, and sometimes competitive collection of women's organizations, persuade them to agree on a course of action despite their other preoccupations, and work simultaneously to influence service providers and policymakers. The health and family planning infrastructure that is projected could serve as a base of operations and influence for women's participation at all levels, and the policy environment for decentralized program implementation is favorable. In the midst of economic crisis and politically volatile structural adjustment policies, however, the implementation of this ambitious infrastructure may prove as elusive as the articulation of a set of reproductive health priorities on which diverse constituencies can agree.

## The Philippines: Confrontations over the right to family planning

The Philippines launched a national policy of population and family planning in 1971 under the direction of the Commission on Population (POPCOM), the chief policymaking, coordinating, and monitoring body of the government's population program. Targeting specific declines in crude birth and death rates and population growth, the Philippines population policy was intended to "make available all acceptable methods of contraception, except abortion," which was illegal,[6] and to expand population education, training, research, and family planning services (Peralta and Ligan, 1975: 62; UNFPA, 1989: 467–478). POPCOM was subsumed under the Ministry of Social Services, whose female minister strongly promoted the family planning program. With substantial support from international donor agencies and the government of the Philippines under the Marcos regime, POPCOM had more than twice the budget of its parent agency.

When 20 years of martial law ended with the "four-day revolution" of February 1986, supporters of democracy hoped for the elimination of the severe economic inequalities and social injustices that had pervaded life in the Philippines for decades. More than two years earlier, women from all classes of society had come together in October 1983 to demonstrate against Ninoy Aquino's assassination and to oppose the martial law regime. Along with other groups and institutions that had formed a multiparty and multifaceted political protest movement—including, significantly, the Roman Catholic Church—women supporting democratization celebrated Corazon Aquino's election to the presidency.

### The first challenges

In the tumultuous months between Aquino's election and the installation of a new legislature, women's rights advocates became embroiled in two major struggles relating to reproductive rights. The first was a battle over the new constitution. Reflecting on their recent history, some participants believe that the women's movement in the Philippines became a specifically feminist movement when activists were confronted by the draft of the national constitution prepared by a 48-member constitutional committee that included only two women. During the fall of 1986, a number of women's organizations, spearheaded by GABRIELA, a national coalition of more than 100 women's groups with almost 40,000 members, prepared a statement on the draft of the constitution to be submitted to the constitutional committee. Concerned with a wide variety of restrictions on their status and rights in the family, at work, and in political life, in addition to the economic and sexual exploitation of Filipinas in low-paid export production, prostitution, and the bar trades, women's groups wanted constitutional protections that would move beyond the simple declaration in Section 14 of the draft that specified "the fundamental equality before the law of women and men."

The women's movement was unprepared for the intense, well-organized campaign managed by the Catholic Church hierarchy, in conjunction with a conservative society of lay Catholics, fundamentalist Christians, and groups such as Pro-Life Philippines, to insert an article into the constitution that would protect the life of the unborn. Women's groups, family planning organizations, and legal experts who favored a more liberal abortion policy held a public seminar in December 1986 and undertook other activities to build public opposition while the draft constitution was under public review. Unable to block the clause entirely, its opponents ultimately succeeded in inserting a guarantee of equal protection of the mother (Marcelo, 1989). Article II, Section 12 of the Philippine constitution, promulgated in 1987, reads, "The State recognizes the sanctity of family life and shall protect and strengthen the family as a basic autonomous social institution. It shall equally protect the life of the mother and the life of the unborn from [the moment of] conception" (Marcelo, 1989: 3).

The second major struggle occurred over a draft of an executive order prepared for Aquino's signature under the auspices of the Roman Catholic Church that would have abolished the Population Commission, restricted the scope of family planning service agencies, banned most contraceptive methods other than natural family planning, and blamed contraception for "fomenting sexually promiscuous attitudes and behaviors" with "perverse consequences" such as "marital infidelity, prostitution and proliferation of sexually transmitted diseases" (quoted in Tadiar, 1989: 90).[7] The executive order would also have repealed all laws or practices supporting incentives for limiting family size, setting population or family-size goals or targets, and promoting or dispensing such "abortifacient contraceptives" as the pill and the IUD. Intense mobilization in February 1987 of women activists and population and health professionals publicized the potential threat to family planning through media coverage, telegrams, and a petition to the president (Marcelo, 1989), and the executive order was never signed. The attacks on family planning were to continue, however, and concerned women's groups mobilized around the protection of their rights.

### Countering ongoing threats

The Aquino government was caught between opposing forces on many complex issues, of which population was only one. It wanted to disassociate itself from the alleged excesses of population control in the Marcos regime and from the implication that rapid population growth was the root cause of poverty. However, several senior officials together with members of the scientific, medical, and internationally funded family planning communities contended that population growth was a critical problem and that family planning must remain a basic right. The National Academy of Science in the Philippines urged that the program be strengthened, not abolished, in order to "improve the general health of the people, decrease child and maternal death rates, and enhance the status of women by opening social and economic opportunities for them" (*Manila Bulletin*, 24 March 1987).

President Aquino was lobbied by all sides. Some nationalist and leftist leaders insisted that the country could accommodate a population twice its current size. Adopting leftist rhetoric, the Church and the "pro-life" movement condemned "coercive and even deceptive" population-control methods and fiercely criticized "foreign meddling" from the United Nations Population Fund (UNFPA), the United States Agency for International Development (USAID), and the World Bank (*Far Eastern Economic Review*, 1988: 25). The new chairwoman of POPCOM was opposed to artificial methods of family planning (*Manila Chronicle*, 21 February 1989). She also opposed continuing the population program, which she insisted publicly had been a failure. Supporters of the national population program countered that blaming the program for the failure to alleviate mass poverty was erroneous. Instead, POPCOM should be moved to a more supportive institutional environment (*Manila Bulletin*, 24 March 1987).[8]

In the midst of the confusion, POPCOM nevertheless prepared a new five-year National Population Plan, according to its mandate and consistent with the new constitution. Approved in principle in April 1987, the plan included a Family Planning and Responsible Parenthood Program intended for married couples of reproductive age with new demographic goals and targets, and an Integrated Population and Development Program. By 1991 the program had not been implemented, however. According to newspaper reports, "An almost three-year struggle for control within the government over the country's population program has culminated in a near-paralysis of a nationwide network of family planning clinics" (*Manila Chronicle*, 21 February 1989: 2; see also 27 July: 1).

Feminists and women's health advocates have organized uncomfortably around the protection of the National Population Program, which is widely viewed as a mechanism for ensuring access to contraceptive services. In 1987, for example, a group called WomanHealth Philippines was formed to bring together women activists and health and family planning professionals from urban and rural areas to promote women's reproductive rights and health (Marcelo, 1989). In addition, GABRIELA set up a Commission on Health and Reproductive Rights to maintain a critical perspective on contraceptive safety and the protection of poor women from coercive practices and lack of informed consent. The WomanHealth coalition made a significant attempt in late 1987 to shape the national population policy when Senator Leticia Shahani, with three colleagues, introduced Resolution 39 to the Philippine Senate urging the strengthening of the National Population Program through the implementation of POPCOM's Policy Statement of April 1987. Shahani, chairwoman of the Senate Committee on Population, had long been an advocate of women's rights at the United Nations. In consultation with the Institute for Social Studies and Action and the Working Group on Health and Population of the Legislative Agenda for Women, among other groups, WomanHealth (1988) proposed extensive amendments to Resolution 39 that would stress the importance of serving women's individual needs and of "humanizing" as well as strengthening the national program. The senate eventually approved a much-diluted version of the resolution in a compromise with the Catholic Church (*Far Eastern Economic Review*, 1988). Legislative attacks on the family planning program intensified, however, with the introduction in 1989 of resolutions that called, first, for research on the efficacy of the Billings ovulation method (natural family planning) as a "universally applicable and acceptable" means of achieving or avoiding pregnancy, and second, for an investigation into the "social costs" of using foreign funds for the national population program (Tadiar, 1989).[9] Feminists found some of their earlier critiques of the population program used against them, as activists opposed to artificial contraception and population control mirrored feminists' language on the dangers and abuses of particular birth control methods and of the population program as a whole (Marcelo, 1989).

The creation of a "Philippine Development Plan for Women" represents a second effort by the women's movement to expand and protect the national population policy and the right to family planning. Formulated by the National Commission on Women with input from women's organizations and ministries as a companion piece to the country's "Medium-Term Philippine Development Plan" for the years 1989 to 1992, the plan specified a wide range of sectoral targets aimed at promoting gender equality. It also reiterated the population targets of the POPCOM policy but added measures to encourage breastfeeding, reduce sexually transmitted diseases, incorporate more comprehensive women's health services into family planning clinics, and—as in Brazil's national women's health policy—to train health and family planning workers to be more sensitive to issues of gender and power (*MARHIA*, January–March 1989). President Aquino moved for approval and adoption of the plan on International Women's Day, 8 March 1989.

According to its spokeswomen, the long-term goals of the women's health movement in the Philippines are to monitor and defeat national legislation that threatens reproductive rights; to empower poor women to demand appropriate, affordable, and accessible health services; to build alliances with key government personnel, researchers, nongovernmental organizations (NGOs), service providers, and legislators to promote high-quality care; and to sustain reproductive health as a public policy issue despite the disarray of the government's population program (*MARHIA*, October–December 1988). The political conflict is intense, however. Continued attacks by conservative forces pose stiff challenges to the feminist health agenda. In addition, the uneasy political alliance with the family planning establishment has been divisive for feminists. Some oppose collaboration with the population-control forces characteristic of the old regime or with any organizations funded by the United States government or US foundations, while others view such alliances as essential for ensuring all women access to a full range of family planning information and services.

## Women's political action in international perspective

In many developing countries, feminist advocates of women's rights and health have taken the initiative in criticizing the inadequacies and excesses of existing family planning programs as well as the ideologies of all hues underlying restrictive antinatalist or pronatalist policies. Their purpose is to expose the biases of prevailing practices and present an alternative view (for example, see Sen and Grown, 1987; Hartmann, 1987; Germain and Ordway, 1989). Feminists view access to safe, acceptable, affordable, and noncoercive contraception and abortion as an essential ingredient of comprehensive reproductive health services, of reproductive rights, and of gender equality for all women.

Feminists have relied on three complementary strategies in their efforts to influence state policy, each of which poses some dilemmas. The first involves building a collective movement that claims to represent the interests of a significant constituency of women across the major social divides of class, caste, ethnicity, religion, and region. The example of Nigeria reveals how difficult this can be in a heterogeneous society where competing identifications obscure a clear awareness of common interests, and in which independent political action has been restricted under authoritarian rule. Nevertheless, some Nigerians do consider themselves part of an international—if not yet a national—women's movement.

Although uneven in its development, the international women's movement has engendered since 1975 a number of associations linking women within and among developing countries. Women who are among the elite meet often at regional or international conferences under the sponsorship of professional associations, governmental and nongovernmental donor organizations, or UN agencies. Some third world networks that are all-encompassing in their political, economic, and social policy agendas, such as DAWN (Development Alternatives with Women for a New Era), based in Rio de Janeiro, have developed critical analyses of reproductive rights and population programs framed in the context of concurrent global issues (Sen and Grown, 1987). Others, such as Isis–Latin America, give more explicit emphasis to reproduction and to women's health. Such global and regional networks of primarily third world feminists stress their roles as researchers, communicators, facilitators, and catalysts for change, as well as their close working relationships with grassroots women's organizations struggling against sexual and class exploitation. Many are affiliated with such international organizations as the Women's Global Network on Reproductive Rights, based in Amsterdam, which links third world groups with individuals and groups in industrialized countries.

Each of these vertical and horizontal alliances within and across countries acts as a resource for exchanging information, articulating policy positions, and planning effective political action within the broad boundaries of the international movement. Exchange and collaboration have not always been easy, however. Most groups are hampered by their lack of experience, contacts, and paid professional staff, and by their slight knowledge of strategies, internal organizational options, and resource-seeking capacity (Staudt, 1981; Germain and Antrobus, 1989). The need for assistance from outsiders, no matter how sympathetic the latter might be, raises additional fears of dependency and control.

A second strategy for influencing policy has required that women who are often unfamiliar with—or suspicious of—the workings of the formal political process acquire political skills (Blay, 1985). The underrepresentation of women (particularly feminists) in key political positions reduces the potential for women's interests to be articulated forcefully. At the same time, activists in the movement, as in Brazil, are often torn between their desire to keep the movement "pure" (and thus to engage primarily in opposition politics) and the desire to institutionalize their programs (and thus to engage in participatory politics, which inevitably requires compromise). Feminists have also found that political achievements can be fragile when the political environment shifts.

The third strategy, building alliances with other institutional actors or constituencies around issues of women's rights and family planning, is also simultaneously both promising and problematic. Most commonly, such allies are individual physicians or organizations of obstetricians and gynecologists, other organized health and family planning providers (for example, nurses and midwives and family planning NGOs), civil rights advocates, social science researchers, key personnel in government ministries, and political parties or candidates seeking a gender-sensitive platform. Divergence of views and mutual misunderstanding can make collaboration difficult, however, or reduce it to a narrowly defined content and perhaps transitory joint action.

Women's health advocates in developing countries have also built alliances with key personnel in international aid organizations working within their countries. The contacts are often initiated by the donor agencies themselves. Virtually all of the major multilateral and bilateral donor organizations and international NGOs in the fields of population, health, family planning, and development now include a few highly committed women and men in their headquarters and field offices who consistently advance—although not always successfully—the cause of gender-sensitive policies and program implementation. Some third world groups, however, have been wary of collaborating with international agencies that they define as representing imperialistic interests or ideologies of population control, even on issues of mutual concern such as reducing maternal mortality. Opponents of collaboration charge that some international aid agencies (as well as national governments) continue to treat third world women in an instrumental manner under the guise of approaching population problems "scientifically" (for example, see Sen and Grown, 1987; Dixon-Mueller, 1993; Hartmann, 1987; Germain and Ordway, 1989). The technocratic approach to policymaking, which relies on "experts" rather than on consultation with the groups that will be most affected by decisions, can be dangerous, critics say, because it gives the appearance of "proposing dispassionate solutions which are devoid of political bias" (Bachrach and Bergman, 1973: 7–8).

Women activists in many countries are responding, not dispassionately, with charges of political bias aimed at the technocrats as well as the politicians. Most governmental and nongovernmental policies and programs in health and family planning have not yet institutionalized the values and priorities of women themselves. In most cases, women's groups are rejecting explicitly pronatalist and antinatalist policies and practices if they threaten to interfere with women's ability to make their own decisions. Focusing on the conditions of women's lives and on the quality of reproductive health services rather than on the achievement of demographic targets, advocates of a "women's perspective" are attempting to transform conventional thinking about the politics and policies of fertility decline in developing countries.

## Notes

This chapter draws in part on field investigations in Brazil, Nigeria, and the Philippines by staff members and consultants of the International Women's Health Coalition: Maggie Bangser, Elizabeth Coit, Joan Dunlop, Adrienne Germain, Andrea Irvin, Janice Jiggins, Deborah Rogow, and Gilberte Vansintejan. Bangser and Rogow also commented extensively on the manuscript, as did Carlos Aramburú, Carmen Barroso, Sônia Corrêa, Mere Kisekka, Kristin Luker, Alexandrina Marcelo, and Jane Ordway.

1 We refer here principally to the existence of autonomous women's organizations and their leadership, rather than to parastatal women's organizations, which serve primarily to reinforce state policies; or to "women's wings" of political parties, religious groups, or protest movements, the actions of which are subject to control by their male counterparts; or to individual women appointed or elected to high public office who remain beholden to their political patrons or to dominant interest groups. Autonomous groups may be elite or mass based, including philanthropic societies; national associations of diverse women's organizations; women mobilized around such issues as peace, social justice, consumer rights, or environmental protection; such occupation-based associations as trade unions, market women's associations, producer cooperatives, or professional associations; rural women's groups; and feminist protest groups.

2 By "explicitly feminist" we mean the recognition that girls and women are subordinated by hierarchies of gender as well as by hierarchies of class, caste, race, ethnicity, nationality, and other criteria. A feminist analysis attempts to understand the sources of female oppression in diverse social systems and to work toward the elimination of such oppression. "Sexual and reproductive rights" refers to the feminist principle that every woman has the right to control her own sexuality and reproduction without discrimination, including the right to regulate her fertility safely and effectively by conceiving only when she wishes, by terminating unwanted pregnancies, and by carrying wanted pregnancies to term (Germain and Ordway, 1989).

3 See, for example, Portugal and Claro (1988), Portugal (1989), and various issues of the *Women's Health Journal*, published in English and Spanish by Isis International and the Latin American and Caribbean Women's Health Network (Santiago, Chile: Isis International). Latin American and Caribbean feminists have been holding large meetings in various countries throughout the region since 1981.

4 The Primary Health Care Program is intended to bring health and family planning services to local populations through 40,000 primary care centers, the first phase covering 52 local government areas in 19 states (World Bank, 1987). These goals are made elusive, however, by user charges levied at all government hospitals under austerity measures, a drop in public expenditures in health care, and inflation. The National Social Development Policy includes a special section on women's welfare with a broad set of goals, including the elimination of customary practices that "dehumanize women" and of religious practices that "militate against their full development" (Kisekka, 1989c: 7). The Commission on Women and Development has been approved (but not yet implemented) to coordinate women's affairs at the national level, with similar commissions planned at the state level. Finally, the Better Life for Rural Women Program was launched by the wife of President Babangida in September 1987 to support women's participation in education, training, agriculture, rural industry, credit, marketing, and health and social services. Initiated from the top, the state and local programs are to be headed by wives of the respective state governors and local administrators. In an effort to deflect possible criticism, the First Lady has been quoted as insisting that "the movement did not support any act of disrespect for the head of the family" (*Woman's World,* 1989: 22).

5 Some researchers and policymakers have urged the government to grant the Specialist Hospitals the right to terminate unwanted pregnancies if and when such a request is made by the individual (for example, see Orubuloye, 1983). Reformists feared, however, that the dominance of USAID funding of family planning services in Nigeria (UNFPA, 1989) would block most women's access to safe services if laws were liberalized, given USAID's prohibitions, until 1993, on the direct or indirect funding of abortion services, counseling, or referrals resulting from the United States' position at the 1984 International Conference on Population in Mexico City (Fox, 1986).

6 Abortion was prohibited absolutely in the 1930 penal code (and in the Spanish code of 1870), although the code was generally interpreted as permitting abortion to save a woman's life. The number of clandestine abortions performed annually has been estimated recently as ranging between 150,000 and 750,000 (quoted in Tadiar, 1989).

7 In its early years, POPCOM included one official representative of the Catholic Church, but the Church withdrew because it was "often defeated in board decisions" (*Manila Bulletin,* 2 February 1987: 9). For a review of the Church's position with regard to the National Population Program under Marcos, see Gorospe, 1976.

8 Ultimately, POPCOM's functions were reduced to program monitoring, while the responsibility for family planning service delivery and coordination of all participating agencies was assigned in July 1988 to the Ministry of Health (*Far Eastern Economic Review,* 1988).

9 Senate Resolutions 286 and 326, respectively. "Pro-life" advocates also introduced additional antiabortion bills in the House of Representatives (HBN 3907, to impose penalties on women who induce or seek abortion) and the Senate (SBN 1109, to increase penalties for professionals who perform abortions, and SBN 1211, to impose the death penalty for the murder of a minor, to include cases of abortion and infanticide). Ironically, the old population-control constituency has also introduced legislation that would grant state benefits to married couples of low income who have only two children within ten years of marriage, would grant incentives for vasectomy or tubal ligation, and would replace POPCOM with a "Department of Population Control."

## References

Alvarez, Sonia E. 1989. "Politicizing gender and engendering democracy," in *Democratizing Brazil: Problems of Transition and Consolidation,* ed. Alfred Stepan. New York: Oxford University Press, pp. 205–251.

Bachrach, Peter and Elihu Bergman. 1973. *Power and Choice: The Formulation of American Population Policy*. Lexington, MA: D.C. Heath.

Barroso, Carmen. 1989. "Maternal mortality: A political question," in *Quando a Paciente é Mulher*. Brasilia: National Council on Women's Rights, pp. 9–12.

———. 1990. "Women's movements, state policies, and health in Brazil," in *Beyond the Decade*, ed. Geertje Lycklama. London: Gower Press.

Basnett, Susan. 1986. *Feminist Experiences: The Women's Movement in Four Cultures*. London: Allen and Unwin.

Blay, Eva Alterman. 1985. "Social movements and women's participation in Brazil," *International Political Science Review* 6, 3: 297–305.

Corrêa, Sônia. 1989. "Reproductive rights in the context of Brazilian demogrphic transition," paper presented at the Annual Meeting of the Population Association of America, Baltimore, March.

Dennis, Carolyne. 1987. "Women and the state in Nigeria: The case of the Federal Military Government 1984–85," in *Women, State, and Ideology: Studies from Africa and Asia*, ed. Haleh Afshar. New York: State University of New York Press, pp. 13–27.

Dixon-Mueller, Ruth. 1993. *Population Policy and Women's Rights: Transforming Reproductive Choice*. Westport, CT: Praeger.

Duchen, Claire. 1986. *Feminism in France from May '68 to Mitterrand*. London: Routledge and Kegan Paul.

Enabulele, Arlene Bene. 1985. "The role of women's associations in Nigeria's development: Social welfare perspective," in *Women in Nigeria Today*, ed. Editorial Committee, Women in Nigeria. London: Zed Books, pp. 187–194.

*Far Eastern Economic Review*. 1988. 20 October, pp. 24–30.

Federal Republic of Nigeria. 1988. *National Policy on Population for Development, Unity, Progress and Self-Reliance*. Lagos: Government Printing Office.

Fox, Gregory H. 1986. "American population policy abroad: The Mexico City abortion funding restrictions," *Journal of International Law and Politics* 18, 2: 609–662.

Germain, Adrienne, and Jane Ordway. 1989. *Population Control and Women's Health: Balancing the Scales*. New York: International Women's Health Coalition, in cooperation with the Overseas Development Council.

———, and Peggy Antrobus. 1989. "New partnerships in reproductive health care," *Populi* 16, 4: 18–29.

Gorospe, Vitaliano R. (ed.). 1976. *Freedom and Philippine Population Control*. Quezon City, Philippines: New Day Publishers.

Hartmann, Betsy. 1987. *Reproductive Rights and Wrongs: The Global Politics of Population Control and Contraceptive Choice*. New York: Harper and Row.

Ilumoka, Adetoun. 1989. "Reproductive rights: A critical appraisal of the law relating to unwanted pregnancies and abortion in Nigeria," paper presented at the conference of the Society of Obstetrics and Gynaecology of Nigeria, Calabar, September.

Instituto de Ação Cultural, Organizado pela Equipe do Projeto-Mulher. No date. *As Mulheres em movimento*. Rio de Janeiro: Editora Marco Zero.

Kisekka, Mere N. 1989a. "Population policy and women's associations in Nigeria," paper presented at the Annual Meeting of the Population Association of America, Baltimore, March.

———. 1989b. "Mapping exercise of women's associations in Jos, Kano, Mubi, and Zaria." New York: International Women's Health Coalition.

———. 1989c. "Reproductive health research and advocacy: Challenges to women's associations in Nigeria," paper presented at the conference of the Society of Obstetrics and Gynaecology of Nigeria, Calabar, September.

Ladipo, Oladipo A. 1989. "Preventing and managing complications of induced abortion in Third World countries," *International Journal of Gynaecology and Obstetrics* Supplement 3: 21–28.

*Manila Bulletin*. 1987. "Population program under scrutiny," 2 February, p. 3.

———. 1987. "Changes likely in national population program," 24 March, p. 1.

*Manila Chronicle*. 1989. "Power struggle leaves family planning network paralyzed," 21 February, p. 1.

———. 1989. "Cautious words throw light on key issue," 27 July, p. 1.

Marcelo, Alexandrina B. 1989. "Reproductive rights and challenges in the Philippines," paper presented at the Annual Meeting of the Population Association of America, Baltimore, March.

MARHIA (Medium for the Advancement and Achievement of Reproductive Rights, Health Information and Advocacy). 1988–89. Quezon City, Philippines: Institute for Social Studies and Action.

Mba, Nina Emma. 1982. *Nigerian Women Mobilized: Women's Political Activity in Southern Nigeria, 1900–1965*. Berkeley, CA: Institute of International Studies, University of California, Berkeley.

Merrick, Thomas W., and Douglas H. Graham. 1979. *Population and Economic Development in Brazil: 1800 to the Present*. Baltimore: Johns Hopkins University Press.

Mies, Maria. 1986. *Patriarchy and Accumulation on a World Scale*. London: Zed Books.

*New Nigerian*. 1989. 18 March, p. 1.

Nichols, Douglas, O. A. Ladipo, John M. Paxman, and E. O. Otolorin. 1986. "Sexual behavior, contraceptive practice, and reproductive health among Nigerian adolescents," *Studies in Family Planning* 17, 2: 100–106.

Nigeria, Federal Office of Statistics. 1992. *Nigeria Demographic and Health Survey*. Lagos: Federal Office of Statistics; Columbia, MD: Institute for Resource Development/Macro International.

Ogunsheye, F. Adetowun et al. (eds.). 1988. *Nigerian Women and Development*. Ibadan, Nigeria: Ibadan University Press.

Okagbue, Isabella. 1990. "Pregnancy termination and the law in Nigeria," *Studies in Family Planning* 21, 4: 197–208.

Okonjo, Kamene. 1983. "Sex roles in Nigerian politics," in *Female and Male in West Africa*, ed. Christine Oppong. London: George Allen and Unwin, pp. 211–222.

Orubuloye, E. I. 1983. "Toward national policy on population," in *Population and Development in Nigeria*, eds. I.O. Orubuloye and O.Y. Oyeneye. Ibadan, Nigeria: Nigerian Institute of Social and Economic Research, pp. 170–178.

Peralta, Ana Maria Rotor, and Marlene C. Ligan. 1975. *Philippine Population: Implications, Program and Policies*. Manila: University of the East Press.

Pinotti, José Aristodemo, and Aníbal Faúndes. 1989. "Unwanted pregnancy: Challenges for health policy," *International Journal of Gynaecology and Obstetrics* Supplement 3: 97–102.

Portugal, Ana Maria (ed). 1989. *Mujeres e Iglesia: Sexualidad y Aborto en America Latina*. Washington, DC: Catholics for a Free Choice and Distribuciones Fontamara, S.A.

———, and Amparo Claro. 1988. "Virgin and martyr," *Conscience: A Newsjournal of Prochoice Catholic Opinion* 9, 4: 14–18.

Rutenberg, Naomi, and Elisabeth A. Feraz. 1988. "Female sterilization and its demographic impact in Brazil," *International Family Planning Perspectives* 14, 2: 61–67.

Sarti, Cynthia. 1989. "The panorama of feminism in Brazil," *New Left Review* 173.

Schmink, Marianne. 1981. "Women in Brazilian abertura politics," *Signs: Journal of Women in Culture and Society* 7, 1: 115–134.

Sen, Gita, and Caren Grown. 1987. *Development, Crises, and Alternative Visions: Third World Women's Perspectives*. New York: Monthly Review Press.

Sherris, Jacqueline D., and Gordon Fox. 1983. "Infertility and sexually transmitted disease: A public health challenge," *Population Reports* Series L, 4. Baltimore: Population Information Program, Johns Hopkins University.

Shrier, Sally (ed.). 1988. *Women's Movements of the World: An International Directory and Reference Guide*. London: Longman.

Staudt, Kathleen A. 1981. "Women's organizations in rural development," in *Invisible Farmers: Women and the Crisis in Agriculture*, ed. Barbara C. Lewis. Washington, DC: Office of Women in Development, USAID, pp. 329–400.

Tadiar, Alfredo Flores. 1989. "Commentary on the law and abortion in the Philippines," *International Journal of Gynaecology and Obstetrics* Supplement 3: 89–92.

Tinker, Irene, and Jane Jaquette. 1987. "UN Decade for Women: Its impact and legacy," *World Development* 15, 3: 419–427.

United Nations, Department of International Economic and Social Affairs. 1988a. *Case Studies in Population Policy: Brazil*, Population Policy Paper No. 17. New York: United Nations.

———. 1988b. *Case Studies in Population Policy: Nigeria*, Population Policy Paper No. 16. New York: United Nations.

United Nations Population Fund (UNFPA). 1989. *Inventory of Population Projects in Developing Countries Around the World 1987/88*. New York: UNFPA.

Wieringa, Saskia (ed.). 1988. *Women's Struggles and Strategies*. Aldershot, Great Britain: Gower Press.

WomanHealth. 1988. "Proposed Amendments to Senate Resolution No. 39 Introduced by Senators Shahani, Tanada, Romulo and Mercado," Manila: WomanHealth, January.

*Woman's World*. 1989. 1 January, p. 22.

Women in Nigeria. 1985. *The WIN Document: Conditions of Women in Nigeria and Policy Recommendations*. Samaru, Zaria: Women in Nigeria.

World Bank. 1987. "Nigeria: First national population project," Initiating Project Brief. Washington, DC: The World Bank, 26 May.

World Health Organization. 1986. *Maternal Morality Rates: A Tabulation of Available Information*. Geneva: Division of Family Health, WHO.

# Limits to Papal Power: Vatican Inaction After *Humanae Vitae*

CHARLES B. KEELY

THE ROMAN CATHOLIC CHURCH is high on most lists of transnational actors that influence the politics of family planning in developing countries. The presumption is that the Church is a counterweight to multilateral and bilateral funders, national bureaucracies, and nongovernmental organizations that promote family planning.

That the Catholic Church is a counterweight is based on two assumptions. The Church is "counter" because it opposes "artificial" birth control. The encyclical *Humanae Vitae*, issued in 1968, reiterated opposition to most birth planning methods at a time when family planning was being firmly etched on global agendas for development and aid. The second assumption is that the Church is a "weight" because of its role in many developing countries and the ability of the Holy See, including the pope and the Vatican Curia, to mobilize resources, including members, to advance its interests and positions.

The thesis of this chapter is that the Holy See has not been the counterweight to family planning that it was expected to be (and sometimes is presumed to have been) from the encyclical's release in 1968 to 1990. That Church teaching opposes most contraceptive practices is true. That successive popes or the Vatican Curia have used their weight to translate those teachings into national or multilateral agency policies is not so clear. Indeed, the lack of vigorous follow-up to *Humanae Vitae* in the political realm requires explanation. Vatican inaction is significant for an understanding of the course of family planning policy in developing countries over the last 25 years.

This chapter reviews the developments leading to the pivotal publication of *Humanae Vitae* and its relation to events outside the Church in those early days of the global spread of the family planning movement. It then explores why the central authorities of Roman Catholicism did not actively pursue a goal of incorporating Church teaching into laws and policies of countries in which the Church was a significant presence and in multilateral organizations in which it has representation.

## The development of *Humanae Vitae* and its aftermath

On 29 July 1968 Pope Paul VI issued his long-awaited encyclical on birth control, *Humanae Vitae*. Pope John XXIII, predecessor of Paul VI, had reserved the issue of contraceptive use to himself rather than have the assembled bishops deliberate formally about birth control or issue any directives at the Second Vatican Council (1962–65). Pope Paul VI, who was elected during the Council, continued this policy.

Although he reserved the question to himself, John XXIII nevertheless appointed seven men to the Pontifical Study Commission on Family, Population, and Birth Problems. The original purpose was to study United Nations policies on population and to advise the pope on appropriate responses in light of Church teaching (Shannon, 1970; on developments within the UN system in the early 1960s, see Symonds and Carder, 1973). The commission first met in Louvain in the fall of 1963, following Pope John's death the previous June. The group proposed that it be enlarged and become more representative. In the spring of 1964, Pope Paul VI instructed the group to focus on moral and doctrinal aspects of birth control. At its third session, specially summoned in May 1964, the pope sought advice from the group (now enlarged to 13 men) on the morality of the oral contraceptive pill, because he planned to make a firm statement on birth control at a consistory (meeting) of the cardinals planned for June. The group failed to develop a consensus (Shannon, 1970).

Paul VI continued the commission and expanded it again to include a broader cross-section of relevant expertise and to include representatives from many countries and married couples. The commission finally comprised 58 bishops, theologians, biomedical and social scientists, physicians and psychiatrists, and male and female leaders of lay family organizations. It delivered its final report to the pope in June 1966. The majority recommended reformulation of the Church's teaching on the legitimacy of using some birth control methods other than periodic abstinence. The contents of the papal commission's confidential report were widely known. They were published in Europe and North America without authorization as early as April 1967, 15 months prior to *Humanae Vitae* (Kaiser, 1985).

The encyclical did not contain the change or development in the Church's teaching on birth control suggested by the papal commission. Instead, it reiterated the ban on contraceptive methods (including hormonal contraceptive pills, intrauterine devices, and barrier methods) that interrupted or altered biological processes so that conception was prevented.

The proscription was founded on the teaching that a necessary connection exists between sexual intercourse, which is to be limited to married partners, and procreation. "Every matrimonial act must remain open to the transmission of life. To destroy even only partially the significance of intercourse and its end is contradictory to the plan of God and to his will. Similarly excluded is any action, which either before, at the moment of, or after sexual intercourse, is specifically intended to prevent procreation —whether as an end or a means" (*Humanae Vitae*, paragraph 14).

*Humanae Vitae* is a clear restatement of opposition to contraceptive methods other than periodic abstinence. It was promulgated by the pope on his own authority after much consultation, but against the majority opinion of an advisory commission and after prolonged personal deliberation by the pope.

In the encyclical, Paul VI appealed to public authorities to prevent enactment of policies and programs that promoted the banned methods. "To rulers, who are principally responsible for the common good, and who can do so much to safeguard moral customs, we say: Do not allow the morality of your peoples to be degraded; do not permit that by legal means practices contrary to the natural and divine law be introduced into that fundamental cell, the family" (*Humanae Vitae*, paragraph 23).

In the past Vatican authorities had attempted and supported attempts to influence family planning policy. In 1960, for example, the bishops of Puerto Rico labeled voting for the Popular Democratic Party of Governor Luis Muñoz Marin a grave sin. To influence the issues of religious education and laws on birth control was the clear objective of a heavy-handed intrusion into electoral politics (Stycos, 1968). As Stycos noted in his account: "When a Vatican spokesman was asked to comment on the Bishops' pastorals, he justified them on only two grounds—that the Bishops were within their authority, and that 'for a number of years the Puerto Rican religious and moral situation has been one of particular gravity. . . it is sufficient to recall the intensive campaign which occurred there in favor of birth control'" (Stycos, 1968: 105).

In their history of the United Nations' population efforts between the end of World War II and 1970, Symonds and Carder (1973: xvi) conclude: "The Roman Catholic Church . . . put pressure on governments to prevent any resolutions or action by international agencies which might encourage the practice of what were regarded as illicit methods of birth control."

Efforts to block the spread of birth control programs were not confined to the public sector. Frank Notestein, influential in the development of the UN Population Division, the Office of Population Research at Princeton University, and the Population Council, noted that it was Francis Cardinal Spellman of New York who "lobbied against Rockefeller Foundation involvement in international population work, and helped to slow the

Rockefeller Foundation's entry into the field" (Donaldson, 1990: 10, citing Notestein, 1971: 79).

The conventional wisdom among members of the population community is that the influence of Catholicism was a principal reason why governmental efforts to slow population growth did not take hold more quickly. The exact mechanism of this influence or causation is often unclear. One could conclude that the authors who generated this perception suspect Vatican coordination, but they often refer only vaguely to the "Catholic Church" or "bishops." Donaldson gives a number of examples of this view. He cites Phyllis Piotrow's charge that "the attack of the Catholic bishops" slowed acceptance of the Draper Committee recommendation in 1959 for US governmental action to reduce population growth (Donaldson, 1990: 10, citing Piotrow, 1973: xii; and see Donaldson, 1990: 23, on the Draper Committee). Donaldson notes that "Malcolm Potts and Peter Selman claim the Catholic church was behind what they [Potts and Selman] call 'a willful act of community self-mutilation' that brought Latin America its population problems . . ." (Donaldson, 1990: 10, citing Potts and Selman, 1979: 311).

Subsequent to *Humanae Vitae*, Church authorities have waded into political turmoil on another moral issue of importance. What has been described as the "political hardball" of abortion debate in North America has been repeated in Europe, not least in the pope's front yard in Italy, as well as in developing countries (Steinfels, 1990). When "reproductive rights" are introduced into family planning policy debates in developing countries, local hierarchies weigh in, as recently occurred in the Philippines and Brazil. (See Dixon-Mueller and Germain, in this volume; and Barroso and Correa, 1990.) One need not fantasize about a Vatican situation room with monsignori plotting political strategy to combat abortion. Given the consistency and vigor of official Church opposition regarding abortion, it would be naive to presume that the Vatican refrains from encouragement, instruction, and even tactical advice delivered via its diplomatic corps around the globe.

Vatican behavior on contraception before *Humanae Vitae*, more recent patterns of behavior in opposition to abortion, and conventional wisdom on the role of the Church in family planning policy provide a reasonable basis from which to expect signs of concerted Vatican effort to have *Humanae Vitae* guide national policies. This expectation is reinforced by the exhortation that Pope Paul addressed to rulers in the encyclical itself, that they "not permit by legal means practices contrary to the natural and divine law" (*Humanae Vitae*, paragraph 23).

Had such an effort been undertaken, it could have had profound effects on the politics of family planning in developing countries. Although the usual precautions about trying to demonstrate a negative are salient, there is no evidence of a concerted effort made directly by the Vatican or orchestrated through local hierarchies to translate the teaching of *Humanae Vitae* into social policy. Symonds and Carder (1973: 200) note that "little effort seems to have been made to impose the doctrine of the encyclical rigidly" after its publication. The noted Jesuit theologian and bioethicist Richard McCormick, in an article on "The Silence Since *Humanae Vitae*" (1974), suggested that Vatican inaction up to that point indicated that a propitious time had come to reconsider papal teaching. The current pope reiterates the teaching, especially on his frequent pastoral visits overseas, but through the period analyzed here (until 1990), direct pressure on national or UN policies, if it existed, must have been exquisitely subtle.[1]

This absence of indicators of Vatican political follow-up does not suggest a lack of effort by individual bishops and various lay or church-sanctioned groups to oppose the introduction and spread of family planning programs. Some people are sincerely convinced that birth control is evil and immoral in itself or because of its effects, whether these effects are described as genocidal, imperialistic, or as class warfare. The opponents of family planning are not and have never been restricted to members of the Catholic Church. The coincidence of Marxist and Vatican positions on family planning at the 1974 World Population Conference in Bucharest has been noted (Finkle and Crane, 1975; Donaldson, 1990; Stycos, 1968). That there is opposition to birth control and family planning by identifiably Catholic sources is neither remarkable nor indicative of concerted Vatican action. The lack of evidence of Vatican coordination (like repeated patterns of political involvement, or consistency in argument and behavior) is notable given behavior prior to *Humanae Vitae* on birth control, more recent behavior on abortion, and the conventional wisdom about Church influence on family planning.

Alongside the conventional wisdom in family planning literature that Catholicism slowed the spread and implementation of family planning,[2] there exists another strain of received wisdom that in the 1960s and 1970s the Catholic hierarchy "made deals." That is, local bishops and their representatives negotiated terms or conditions under which the local Church authorities would not oppose or interfere with government-sponsored family planning programs. The geographic breadth of reported instances of such deals and the known loyalty to Rome of the hierarchies involved would lead to the conclusion that the Vatican at least tacitly assented to this course of action. Leona Baumgartner, while the New York City Health Commissioner from 1954 to 1962, reportedly agreed not to encourage Catholic women to use contraceptives and, at the same time, Cardinal Spellman did not oppose making family planning services generally available (Donaldson, 1990). Warwick (1982) reports similar understandings in Mexico and the Philippines that came about as a result of national leaders' careful political appraisals of the Church's position, appraisals undertaken while the leaders were implementing national programs. Stycos (1968) makes similar references to Colombia specifically and to Latin America generally. Beyond such instances, the lack of concerted and coordinated action aimed at having the encyclical's teaching and exhortations translated into policy requires explanation.

## Developments in family planning as a background to *Humanae Vitae*

The encyclical that was meant to settle the matter of birth control methods, at least for Catholics, did no such thing. Pope Paul VI concluded that he alone would decide the issue, and he deliberated long over his decision. He opposed his own commission's majority recommendation, and had his final word in the encyclical. But as Cardinal John Sheehan, Archbishop of Westminster, said at the time: "It's too late" (Kaiser, 1985: 5).

By 1968 public opinion polls indicated that a preponderance of Catholic couples in the industrialized world practiced contraception and did so in good conscience (Kaiser, 1985). In Latin America studies from the UN demographic center in Chile (CELADE) indicated that a majority of women in Chile, Peru, and the Caribbean by 1965 had already tried some method of contraception, and similar data reported in 1968 indicated that between 38 percent and 65 percent of urban women in Bogotá, Caracas, Mexico City, Panama City, Rio de Janeiro, and San José had practiced contraception (Stycos, 1968).

Stycos reports (1968: 308) that a Pan American Assembly conference held in Cali, Colombia in 1965 concluded, among other things, that:

> (1) All nations should develop population policies as an integral part of their policies on economic development.
>
> (2) Governments should foster responsible paternity by encouraging couples to have the number of children consistent with their own ideals.
>
> (3) Governments should make family planning services accessible to the people who desire them, and educate the people to their availability.

Stycos's commentary is telling: "These remarkable statements . . . would have been unlikely in 1964, implausible in 1960, and unthinkable in 1955" (Stycos, 1968: 308). This was taking place while the pope and his commission were deliberating.

In addition, institutional developments important for the family planning movement were also taking place while *Humanae Vitae* was being developed. During the 1960s at the United Nations, governments debated involving the organization in the population issue and, specifically, providing technical assistance in family planning under UN auspices. Indeed, these debates prompted John XXIII to establish his birth control commission in 1963 (Shannon, 1970).

"Several events occurred in the months preceding the 20th session of the General Assembly which led many observers to refer to 1965 as the turning point in the history of the involvement of the United Nations in population action programmes" (Symonds and Carder, 1973: 140). These included policy changes by the United States, the UN compliance with India's request to field an expert mission on that country's family planning program, the resolution of the Commission on Women concerning access of married couples to contraceptive information, and the Population Commission's discussion of a long-range program including assistance to family planning programs.

In the mid-1960s the United States began its population program as part of foreign assistance implemented through the Agency for International Development (USAID). "By January 1968, . . . [n]o longer simply the controversial interest of a few private citizens, the effort to control population growth in the developing world had become official U.S. government policy, well funded by Congress, accepted by the public . . ." (Donaldson, 1990: 43). In 1968 the United Nations Fund for Population Activities (UNFPA) was launched, with the first $1 million contributed by the United States. These two agencies, USAID's Population Program and UNFPA, have been major external donors and actors in helping to bring about a revolution in human behavior that a great proportion of the world's population has undertaken in less than a quarter century.

In 1968 the UN declared the ability to determine the number and spacing of children to be a human right (Donaldson, 1990). By the time Pope Paul acted in July 1968, the train had left the station.

## Reactions to *Humanae Vitae*

Against this background, the reaction to *Humanae Vitae* within the Catholic Church (and among many sympathetic observers outside the Church) becomes understandable, even while it is noteworthy. For the most part, the reaction reported in the media was negative and critical (Stycos, 1971; Kaiser, 1985). At the Vatican, the storm of protest was beyond what was expected (Shannon, 1970). The encyclical generated anger and opposition, and not just outside the Church. "The reaction by Roman Catholics to Pope Paul's encyclical is so strong that Protestants have no need to overreact to it," concluded the editor of *The Christian Century* (Shannon, 1970: vii).

The reactions of bishops were unprecedented. At the invitation of the Vatican Secretary of State, many national bishops' conferences felt compelled to issue statements. The request of the Vatican Secretary of State was "to stand firm with the Pope, to present 'this delicate point of the church's teaching' and 'to explain and justify the reasons for it'" (Shannon, 1970: 139). By 1981, statements by 15 national conferences of bishops were issued (Hoyt, 1968; Kaiser, 1985; Shannon, 1970; and Carlen, 1981c).

Aside from their substance, two aspects of these episcopal commentaries are noteworthy. First, although the pope had spoken, bishops were invited to comment and they responded. If the pope had the final word, why should bishops issue commentaries? Second, national conferences of bishops issued these reports. The national episcopal conference was an innovation that has grown greatly since the Second Vatican Council. The theological justification for and the organizational integration of these new groupings were in their infancy at the time and are evolving to this day (Reese, 1989). Some interpreted the collective episcopal responses as an example of the principle of collegiality operating in the Church (Shannon, 1970). That principle, as presented at the Second Vatican Council, asserts that "the bishops constitute a stable body or 'college,' and this college of bishops, with the Pope as its head, is collectively responsible for the tasks of the whole Church" (Shannon, 1970: 117). The Council, however, did not define how the principle of collegiality should operate.

There was a range of opinion in the episcopal commentaries on the pope's encyclical. Some interpreted and some even demurred in the sense that they claimed that the primacy of personal conscience was not in dispute. Although no episcopal conference opposed the ban on contraceptives, many focused on pastoral responses and took pains to underscore that people must follow their conscience, giving attention to the pope's teaching. This support was less than enthusiastic.

At the time of the encyclical many others spoke out, including theologians, members of the papal commission, individual bishops, and a variety of other commentators, on the philosophical argument (natural law) and social science arguments marshaled by the pope for his case against "artificial" birth control. The content of the opposition within the Church to the encyclical was as important as its vigor. Analysis of the opposition permits insight into the organizational dynamics of the Church and what was at stake beyond the substantive issues of *Humanae Vitae*.

The literature on *Humanae Vitae* is, indeed, large. One reference work published in 1981 (Carlen, 1981c) cites 152 commentaries on the encyclical. Many of the early commentaries dwelt on the reasoning in Pope Paul's text. Soon, however, the issues of the binding nature of encyclicals on the conscience of Catholics, the meaning of *magisterium*, or teaching authority, the authority of particular types of Church documents, and even the very nature of authority and its exercise in the Church were examined.[3]

A major controversy among Catholics following publication of the encyclical was how binding it was. Much discussion centered around the magisterium, a word that came into vogue in the nineteenth century to refer to authoritative teaching of the Church and especially the hierarchy.[4]

Two issues arose. First, could the magisterium change? Even members of the papal birth control commission discussed whether prior papal pronouncements could be changed. The encyclical *Casti Connubii*, promulgated in 1930 by Pope Pius XI, condemned contraception as "an offense against the law of God and of nature" and, therefore, a "grave sin" (Kaiser, 1985: 4; Carlen, 1981b: 391–414). How, then, could a new pope permit what was declared a sin against nature? Could the Church have misled people for so many years by imposing the discipline of periodic or complete sexual abstinence on married couples and still remain credible to its followers? Were all those who failed to obey guilty of grave sin and perhaps even condemned to an afterlife of punishment for actions that were now no longer to be judged moral failures?

The theologians on the papal commission sharply disagreed on the answers to these questions. That doctrine could develop with deeper understanding, insight, and even technical knowledge was, for some, the way ethics, moral theology, and Church teachings worked in practice. To others, this idea of development was theological nonsense. In the nineteenth century, Pope Pius IX in his encyclical *Quanta Cura* condemned the "erroneous opinion, . . . that 'liberty of conscience and worship is each man's right . . .'" (Carlen, 1981a: 382). Despite this prior papal teaching, the assembled bishops in the Second Vatican Council took the opposite view. Past Church teachings on usury and geocentric astronomy can be cited as examples of previous magisterial teachings that have been modified.

A second issue, more organizational than theological, was evident. To some in the Vatican Curia and to their theological supporters on the papal birth control commission, teaching could not be both authoritative and changeable. To accept such a proposition would be organizational suicide. For the supporters of the idea of doctrinal development, by contrast, the Church leadership could give guidance bolstered by the best available evidence and thinking. Thus, in the light of progressive knowledge or new insight, the magisterium could develop and still be authoritative as a guide to the formation of Catholics' beliefs and moral conscience. Authoritative was not necessarily synonymous with unchanging or infallible. The proponents of accepting a new stance toward contraception emphasized doctrinal development; the opponents maintained that truth cannot change.

The organizational issue, then, can be rephrased: Was a papal encyclical binding on Catholics; did it partake of the infallibility claimed when the pope alone or in a council with bishops invoked what Catholic doctrine teaches is a divine authority on matters of faith or morals? The technical answer is that *Humanae Vitae* was not presented as an infallible statement of morals by the pope. The question, however, strikes at the very heart of Roman Catholicism as an organization. At issue is nothing less than the structure and authority of the Church. Was authority to be centralized in the office of the pope or was it to be more collegial, that is, more diffuse? Further, what was to be the relationship between ordinary papal or hierarchical teaching, the magisterium, and individual conscience? This latter question contained echoes of the Protestant Reformation.

Kaiser (1985), in his review of events leading up to and following the issuance of *Humanae Vitae*, stresses the importance that continuity of teaching assumed in Vatican circles after the papal commission delivered its report. The members of the commission who opposed change in teaching as recommended by the commission majority worked, he claims, with Vatican colleagues to defeat such change. *Humanae Vitae* was needed not only to make the substantive points about contraception, but also to reaffirm continuity in teaching and the role and authority of the papacy in Catholicism.

The debate within Catholicism following the encyclical was not so much about sex as about power.[5] Rome had spoken. The problem was not that nobody listened. Even worse, those who listened disagreed and they questioned the process, the authority, and the relevance of a papal pronouncement. From an organizational perspective, the stakes were much higher than the degree of digression from teaching on sexual ethics. That Vatican curialists were alarmed is not surprising. They were already doubtful about the effects of the Second Vatican Council on decentralizing authority (collegiality) and the apparently lessening grip of Church direction (at all levels—papal, episcopal, and even the parish) on the attitudes, behaviors, and beliefs of Catholics.

As one author wrote during the early period of protest, "Beyond question it [the resistance to an unambiguous papal ruling] has brought the Roman Catholic church to a state of crisis, from which it can hardly emerge without undergoing profound change" (Hoyt, 1968: 8). Some thought collegiality would be given a boost, as the episcopal commentaries suggested. Others hoped and even called for a reformulation in light of the debate that followed the publication of *Humanae Vitae* (McCormick, 1974).

Rather than increase collegiality or change papal teaching, a pope could attempt to recentralize authority in the Holy See. Such an effort would require reestablishing organizational discipline and unity in teaching, behavior, and action, as well as reasserting the universal pastoral responsibility of the pope that the newfound passion for episcopal collegiality had overshadowed. The spread of birth control was not the only foundation for the desire to reestablish papal priority in practice. Declining vocations and clerical defections, decisions on marriage annulments, experiments in liturgy, innovations in theology concerning the nature of the Church (ecclesiology) and morals (premarital sex, homosexuality, divorce, abortion, experiments to overcome infertility including artificial insemination and in vitro fertilization)—all had to be addressed. These issues were matters of concern to those clerical and lay Church members who thought that the Second Vatican

Council had been distorted by liberal theologians into a license to change. Many thought developments had gone too far and often in wrong directions. The time had come to rein things in.

The Roman Catholic Church, however, has multiple and complex goals and tremendous internal variety. As a transnational organization, its leadership, especially the traditional central leadership of curial officials in the Vatican who serve the Holy See, had to survey the Church's position and interests.

## Roman Catholicism in the late twentieth century

If essays were written at five-year intervals describing the structural characteristics of the Roman Catholic Church over the last quarter century, the relative emphasis given to centralized authority as opposed to more diffuse initiatives would be a major point of contrast. Vallier's essay in a collection on transnational actors (Keohane and Nye, 1972) presents the "Catholic Church System" as a centralized monarchy. He writes, "Generalized control at the center is a prerequisite for all other activities" (Vallier, 1972: 131) and goes on to specify this general characterization:

> In its central goals and corporate tradition the Roman Catholic church is first, last, and perhaps always a religious organization. As such its main activities and programs are focused on propagating sacred truths, consolidating religious loyalties and commitments, and creating and renewing sacred structures. In pursuing these objectives and in claiming *for* itself the role of religious leadership the church automatically assigns itself a role as a "moral authority". . . . Moreover the church has fashioned a transnational organization that provides it with a visible center from which its moral principles can be communicated and through which it attempts to exercise influence and social control. (p. 133)

Vallier notes that the Church was undergoing change in the 1960s because of initiatives made by Pope John XXIII and progressives in the Church's hierarchy at the Second Vatican Council (p. 134).

These efforts were part of older trends, Vallier holds, that made the Church more dependent upon spiritual and moral leadership than upon organized political support for its influence. To make itself more credible, the Church's center for the last century yielded some of its autonomy and flexibility to allow local churches and specialized Church agencies (for example, religious orders) to articulate and respond to local situations. The center, progressively less dependent on formal political entanglements, had "more freedom to assert general stands on controversial issues and ethical problems . . . toward a systematic position as global pastor" (Vallier, 1972:

150). These general trends are exemplified by the social encyclicals, starting with Pope Leo XIII's encyclical of 1891 on industrialism, through more recent letters on economic development and the problems of capitalist and communist political and economic systems. The Catholic Action movements in several countries, the worker–priest movement in France after World War II, and the more recent development of base communities as expressions of liberation theology in Latin America are all examples of localized autonomy based on grassroots initiatives. These local initiatives, however, have in some instances resulted in tension between the center and local churches. Rome suppressed the worker–priest movement and the tension between the Vatican and Latin American liberation theologians is widely known.

In a more formal sense, the three main "activities and programs" identified by Vallier, namely, propagating sacred truths, consolidating religious loyalties, and renewing sacred structures, are not necessarily mutually reinforcing. Propagating sacred truths, for example, may not help consolidate religious loyalty. The storm of protest and debate by Catholics around *Humanae Vitae* indicates the potential for dispute over organizational goals.

The relative share of authority and initiative between center and localities in the Roman Catholic Church may shift over time. A secular trend toward devolution of authority and initiative from center to periphery is not irreversible. The global pastoral role that the Holy See can play contains the seeds of a recentralization as the Vatican exercises doctrinal and moral guidance. The current pope, John Paul II, inaugurated his papacy at a ceremony initiating his service as "universal pastor of the church," a new papal title for an old idea (Kelly, 1986: 327). This ceremony was substituted for the former pontifical coronation. While the trappings of monarchy were abandoned, the tendency toward autocracy was reasserted.

Paul had exercised a global pastoral office when he spoke on birth control. What obstacles did the pope and curial officials face in attempting to institutionalize that teaching in national policies and the practice of multilateral organizations, either by direct action of the Vatican Curia and its diplomatic corps or through local bishops?

Recalling Vallier's description of the Church's main functions, the Vatican has to think about evangelizing, maintaining loyalty, and keeping the Church's structure operational. In broad strokes the Church's situation in the period since *Humanae Vitae* in major areas of the globe can be summarized.

In the industrialized countries of Europe, North America, and Oceania, the Vatican's major task has been retention of loyalty. These countries' conferences of bishops were the least likely to reiterate the contents of the encyclical in their statements. (The United States bishops' conference was the main exception.) Lesser clergy and lay members in these countries were less likely than their bishops to accept the teaching. The traditional role of the laity in the Church ("pray, pay, and obey") had become outmoded for most local churches. Clearly, birth control practice was far different from Church teaching; Catholics' family planning practices were not discernibly different from those of their non-Catholic neighbors. Church initiatives in the political sphere were unwelcome in most of these countries, because of traditions of separation of church and state, anticlericalism, and religious pluralism. Finally, Catholics in these countries were the most vocal in their opposition to *Humanae Vitae*. The development of that opposition—from questioning the teaching to criticism of the structure that produced it and rejection of the Vatican's traditional claims to authority and to the manner of exercising it—was cause for Vatican concern. Given birth control practice and the weakening loyalty to the papacy, what would the Church gain by trying to outlaw or circumscribe access to birth control? If anything, Vatican political success in this instance, an unlikely outcome in these countries, might weaken loyalty to the Church and lead to broad civil disobedience to laws inhibiting birth control.[6]

In Asia, the Church was confronted for the most part with non-Catholic countries, many of which had clear birth control policies and programs, and the Church's prospects for changing these policies was minimal. In some countries (China, North Vietnam, and North Korea), the major item on the Church's agenda was obtaining a diplomatic understanding that would allow the Church to function. The Philippines was the single Asian Catholic country. There, strong government policy toward family planning in the early 1970s led to one of the "deals" mentioned earlier. Abortion and sterilization were to be officially discouraged and voluntarism was to be safeguarded, but in the end the family planning program would exist without constant attack from the Church's hierarchy.

In Africa, birth control was not widely adopted. The bigger problem for the Church regarding sexual ethics, marriage, and the family was the divergent African view of marriage, seen more as a process than as a single contractual act establishing a relationship. Roman Catholicism perennially faces the problem of enculturation, that is, finding what in a local culture is compatible with the Church's central doctrine—what can be incorporated and what must be challenged as unacceptable.[7] The traditions surrounding African marriage patterns are one of the greatest problems of enculturation for Catholicism. Indeed, Church authorities avoid using the word enculturation, probably because it seems to bring into question the universality of the Catholic message or to smack of a Eurocentered cultural imperialism.

Africa also presents a fertile field for evangelization. Islam is the major competitor, and Islam is no proponent of birth control. Even though learned Muslim leaders permit birth control, on the local level the practice has not been popular. Generally, African societies are pronatalist. There is a difference of opinion among the family planning community about whether African social structure and economically linked gender and family roles will inhibit the spread of a pro–family planning ideology and slow the increase in contraceptive prevalence.

The African situation seemed compatible with *Humanae Vitae*'s ban on modern birth control methods. Because family planning is not widely accepted by the peoples of Africa, despite recent government acceptance in some countries, the Church did not need to focus on birth control as a public policy issue. Instead, the Church sought to evangelize and to emphasize aspects of African culture that are true to Catholic teaching.

In Latin America family planning was gaining ground and contraceptive use was increasing. While loyalty and commitment to the Christian message was growing, loyalty to the Roman Catholic Church, as traditionally understood, was under fire. A major problem for the Vatican in this world region may be expressed as dealing with the excesses of success. As Vallier pointed out, "The most salient 'other' in the environment of the Roman Catholic church during the past 50 years has been communism . . . equally committed to a body of religious truths and equally aspiring to transnational dominance . . . " (Vallier, 1972: 141).

The political left challenged nominally Catholic societies, in which Church officials often were intertwined with oppressive elites. In the 1950s, local churches were "short on personnel, funds, action models, and laymen who could be mobilized to extend the church's influence to new groups and marginal strata. Under these conditions . . . the transnational church, in close cooperation with local hierarchies, swung into action" (Vallier, 1972: 142). Missionaries, funds, and institution-building activities were transferred from Europe and North America. But the major threat from the left was the perceived need of radical social change. What developed from local resources were "concepts and norms that could tie Catholic loyalties to the need for social change" (Vallier, 1972: 143). Liberation theology included an institutionalized mechanism, the base Christian community, which sought to read scripture in a manner that would inform and coincide with political, economic, and social activity. Peasants, the urban poor, students, and intellectuals were targeted—the same groups targeted by the left.

For the Catholic Church's center, these events, while lending credibility to its religious message, contained both conceptual and institutional dangers. Some Church authorities found liberation theology too materialistic. Marx seemed to be converting the theologians. In addition, some Catholics applied the liberation analysis of social institutions to the Church itself, labeling the latter an oppressor.

Within the institution of the Church, base communities encouraged members to apply scriptural insights to current events and social structures. Lay members took active roles in leadership of the communities. The Vatican felt that these developments not only reduced the teaching and leadership role of the clergy and hierarchy, but also could lead to error. The sincerity and dedication of the laity were not in question, but the maintenance and re-creation of the sacred structure, the Church, were implicitly threatened. Specifically, the divinely ordained offices meant to guard the truth of the message given by its founder were devalued. It was the hierarchy's role (with the aid of the clergy) to teach and to preach.

A more recent worry for the Catholic Church in Latin America is the number of conversions to fundamentalist and Pentecostal Christian sects (Stoll, 1990). Protestant resources in the form of funds and missionaries, emanating especially from the United States, were a factor in the inroads made by fundamentalist Christians. This development could be interpreted in two ways. First, it indicated the need for something like base communities in Catholicism, if the community aspect of religion was to be effective in the lives of Latin Americans. The Protestant communitarian sects seemed to respond to needs and tap into wells that were similar to those of the base communities. Alternatively, Protestant sects could be seen as another expression of the dangers of unchecked enthusiasm. The need for discipline and authority to guard the truth and to test and validate teaching was underscored by these deviations from sound practice.

In summary, if the central Catholic leadership surveyed the Church's situation across the continents over the last two decades, the challenges and preoccupations that confronted them would not dictate vigorous central action to influence birth control legislation, policy, or programs. In fact, as early as 1974, an alternative strategy on population issues was articulated for the Church that focused on papal teachings concerning social development rather than the morality of contraceptive methods.

One such expression was the essay by Hehir (1974) on the eve of the UN Population Conference in Bucharest. He proposed that the Church emphasize the issues of consumption and the distribution of wealth and resources in the international system. Hehir pointed out that in the social encyclical *Populorum Progressio*, issued 16 months prior to *Humanae Vitae*, Paul VI explicitly affirmed an objective problem of population growth as a legitimate area of government intervention, but without specifying its scope (Carlen, 1981c). He distinguished between personal morality and public policy: That papal teaching prohibits contraception except for periodic abstinence does not mean that pressures ought to be applied to governments to eschew contraceptive techniques as part of their interventions to address rapid population growth.

Hehir's article is one example of an intellectual justification for the Vatican to refrain from coordinating a program designed to affect family planning programs. The Church's lack of a basis from which to take direct action, the "prudential calculations about the consequences for the Church and society of seeking to make the teaching [of *Humanae Vitae*] the norm of societal action" (Hehir, 1974: 76), the initial storm of opposition, and the serious questioning of the institutional expression of Roman Catholicism that evolved from the commentaries on the encyclical, all provide the explanation for Vatican inaction.

The Bucharest Conference in 1974 was hardly an endorsement for vigorous international activity to promote family planning. The Indian delegation's statement that "Development is the best contraceptive" was a slap in the face to family planners (Finkle and Crane, 1975). Vatican action therefore was unnecessary.

A sequence of events can be outlined that account for Vatican behavior since *Humanae Vitae*, from 1968 to 1990, from Paul VI to John Paul II.

## Vatican inaction after *Humanae Vitae*

No single hypothesis put forward to explain the apparent absence of coordinated Vatican action to stymie family planning around the world will do justice to the events of more than 22 years under three popes. Although the Vatican does not act openly, the contexts and events described provide the basis for an explanation of the behavior of central Church authorities.[8]

The immediate reaction to *Humanae Vitae*, the storm of protest, took the Vatican by surprise. Paul VI acknowledged this only a month after his encyclical when he expressed his hope that greater understanding would result from the "lively debate aroused by our encyclical" (Shannon, 1970: iii). The Vatican Secretary of State's unusual call for statements by bishops' conferences was an appeal to bolster a shaken papacy by a show of unity.

Paul's reaction, to let the lively debate develop, was in line with his reputation for deliberation. He was sometimes faulted for this character trait. Pope John XXIII referred to him as Hamlet-like (Kelly, 1986: 324). Vatican wags used to describe the "Pope Paul dance": one step forward, one step backward, one step left, one step right; no matter how often repeated or how vigorously engaged in, you always end up in the same place. The application to *Humanae Vitae* is clear.

The encyclical contained gentler language than its predecessor, *Casti Connubii*, which condemned birth control in 1930. The argumentation in *Humanae Vitae*, however, was generally criticized as weak. One majority member of the papal commission noted that if the pope was intent on reaffirming the ban, the commission could have written a better encyclical. As others had done, the commission member faulted the biological reductionism of a presumed connection in natural law between each act of intercourse and a requirement to be open to conception, as well as the empirically unsupported statements about contraceptive availability leading to promiscuity and husbands' increased tendency to treat their wives as sex objects.[9] Only three years before the encyclical, the UN Social Commission was equally confident that "there was agreement that national family planning programmes were essential in strengthening family life and improving the status of women" (Symonds and Carder, 1973: 141).

Vatican political inactivity during Paul VI's pontificate (through 1978) is reasonably explained by the deliberate response of the pope to the protest brought on by the encyclical. That he let the issue develop and the lively debate continue without vigorous intervention was characteristic. Many curialists and conservative Church leaders felt that Paul let too many things drift. Paul seemed bent on adding to the confusion by appointing men of both a progressive and a conservative cast as bishops.

On the other hand, Paul's patience after *Humanae Vitae* may have been farsighted. When he had not acted decisively before 1968, "couples by the millions stopped waiting" (Hoyt, 1968: 10). Any vigorous attempt after *Humanae Vitae* to impose Church teaching had limited possibilities and many pitfalls, as the review of prospects for evangelism, retention of loyalty, and structural renewal indicated. There were many hazards from probable backlashes to an interventionist strategy. There was a consistent rationale and justification for nonintrusion into political matters, as exemplified by Hehir's argument. The Bucharest Conference in 1974 showed that the developing world was not yet ready to embrace family planning, as third world leaders dealt a blow to what was presumed would be the endorsement of an American-arranged agenda.

While Paul waited, theologians also "kept thinking—and not only about birth control, but also about infallibility, about the 'new morality,' about the role of the church in guiding consciences, and about the validity of the traditional modes by which the church seeks the truth" (Hoyt, 1986: 10).

In 1978 two new popes were elected. The first, John Paul I, died shortly after election, to be replaced by the cardinal-archbishop of Krakow, the first non-Italian pope since 1522. John Paul II had the reputation of being a doctrinal conservative, a social liberal, and a wily tactician who had dealt with communist governments successfully to maintain a vibrant, traditional Catholicism in Poland.

Where Paul practiced apparently prudent restraint, John Paul seemed to perceive dangerous drift. The new Polish pope set a number of projects for himself. He wanted to open up the communist East to Christian practice. A traditional function of the papacy is to ensure, through diplomacy, conditions for the free exercise of the Catholic religion (Vallier, 1972). Another project was evangelism, the subject of an encyclical early in his pontificate. A third was trying to bridge longstanding rifts among Christians. Most significant for the present analysis is that Pope John Paul has chosen to reassert papal authority.

The pope has attempted to accomplish this last project in a number of ways. First, he appointed bishops who are Vatican-centered in their view of the Church and whose opinions coincide with his own in doctrinal and organizational matters (McBrien, 1990). Second, he asserted a global pastoral office for the pope as the visible head of the church. While not denying collegiality, he has swung the pendulum back toward a monarchical model for the Church in which the pope is in charge and bishops are akin to viceroys carrying out policy. There is a tension between the office of the pope and the principle of collegiality, as they have developed historically. John Paul's theology of the Church clearly emphasizes his obligations and prerogatives. Third, he has imposed discipline on Church officials, the clergy, and religious orders. Fourth, he has challenged theological dissent. The cardinal in charge of the Congregation of the Doctrine of the Faith, Joseph Ratzinger, is widely regarded as both stalking horse and lightning rod for the pope's views.[10] Cardinal Ratzinger articulates the conservative view and imposes discipline on theologians. The attempt to silence theologians and the recent call for bishops and theologians to refrain from any public dispute with Church teaching or practice are part of a trend under John Paul to mold a Church that speaks with one voice.

Pope John Paul has practiced a modified version of the policy of benign neglect on the contraceptive issue. He has not vigorously attempted to influence national policies. Even the bishops he has appointed have not engaged in concerted political pressure on birth control, at least through 1990. The pope has not abandoned the teaching in the encyclicals of Pius XI and Paul VI. He repeats it frequently and predictably in sermons and discourses on his many trips, billed as pastoral journeys. He apparently approves of attempts to quiet theologians' dissent about the Church's teaching.

John Paul's priority seems to be the reestablishment of the Church structure, which he sees as essential to a healthy Catholicism. To have taken on the issue of birth control frontally would have been imprudent. Rome-oriented (and, therefore, loyal) bishops had to be put in place. Dissent, especially by theologians, had to be quelled. As the Church approaches a situation of one truth spoken by one voice, various issues can be addressed. Abortion is a more serious moral threat to the Church than contraception and has engaged the Vatican and local hierarchies around the world far more than birth control.

The silence after *Humanae Vitae*, from 1968 to 1990, can be summed up as having three stages. First, there was stunned silence, followed during the rest of Paul VI's pontificate by a strategy of restraint while the forces at work—on that and many other ecclesiastical, social, and moral issues—labored in a decade of extraordinary ferment. The final stage has been a tactical silence while internal order is being restored by the reestablishment of centralized authority in the person of the pope and derivatively in the Vatican Curia.

## Conclusion

This analysis follows from the presumption that had the Vatican attempted to translate teaching in *Humanae Vitae* into action, the history of family planning in the last quarter century could have been altered. In fact, Vatican inaction has been important to the politics of family planning in developing countries.

One lesson of this essay is that the idea that power in Roman Catholicism is static should be abandoned. The notion of a fixed authoritarianism able to command great numbers into concerted action is inaccurate. Following Paul VI's inaction, John Paul II has attempted to reassemble papal power through his view of theological doctrine, his trips, his powers of episcopal appointment, and the discipline of canon law. The pope has a considerable number of clerical and lay allies who agree with his vision, his analysis of what needs doing, and his project for reasserting papal authority. Whatever the current situation, it is clear that the central power of the Church can vary over time.

The power of the Church in international affairs and public policy is not confined to papal centralism. The transnational voluntary association called the Roman Catholic Church is characterized by multiple sources of influence and power rooted in its transnational character. The office of the papacy and the dynamics of the relationship of the papacy to national hierarchies and individual bishops is one such source, and a major one. Perhaps too much emphasis is given to this source. Both the central and the local influence of Catholicism vary over time; they can differ from issue to issue and they are related to geography. Church influence on local issues can be extremely important without Vatican initiative or direction.

The reaction to *Humanae Vitae* should lead to a reevaluation of the conventional wisdom about Roman Catholicism's role as an international actor. An appreciation of why the Vatican did not play a more influential role in the family planning policies of countries and multilateral organizations (as well as in the behavior of Catholics) may lead to a better understanding of the Church's organizational dynamics, the limits to papal power, and the actions of the Holy See and the Vatican Curia on this and other issues in the international arena.

## Notes

1 Vatican opposition to abortion in multilateral settings and the apparently concerted activities on national levels regarding legalization of abortion are cited by some as indicative also of activity regarding contraceptive practice. (See, for example, Dixon-Mueller and Germain in this volume and Barroso and Correa, 1990.) In the cases referred to (often Brazil and the Philippines), the issue has been joined over reproductive rights. That phrase is often synonymous with legalizing abortion. In that context, local hierarchies assume that prior agreements or understandings between governments and Church officials regarding the scope and content of state family planning are under attack. (See text for a discussion of such "deals.") Raising the issue of contraception is more a reactive tactic than an indication of a concerted offensive to insert Church teaching about contraception into national law. A firmer indication of a new Vatican effort has taken place sub-

sequent to the period analyzed here and also subsequent to the drafting of this chapter. In the summer of 1992, the Holy See successfully led an effort to strike references to family planning from paragraph 99 of Agenda 21 of the UN Conference on Environment and Development at Rio de Janeiro. At the executive board meeting of the United Nations Children's Fund, over the strong objections of the Holy See, the board asked the agency to cooperate with the United Nations Population Fund and the World Health Organization "to support family planning in the context of sustainable national health systems" (Deen, 1992: 3–4). These activities indicate a probing on the part of the Vatican within the relative safety of multilateral organizations that are a step removed from governments. The strong opposition to and the mixed success of these activities suggest that further efforts, particularly on the country level, may be counterproductive. Whether they will be followed by efforts directed at national policies remains to be seen.

2 A dissenting literature exists on this conventional wisdom—for example, Day, 1967 and Miró, 1963, cited in Stycos, 1968.

3 I consulted some of this literature for this article but it is a different level of discourse from the analysis of texts, arguments, and actions that the social scientist undertakes in order to discern policy, its development, and its consequences. Matters of theology form part of what is to be analyzed, including the notion of some theologies that social science analysis of theological or ecclesiological matters is illegitimate and fundamentally misleading. Such a position is, of course, part of some Churchmen's tactics for harnessing and enhancing their power.

4 To some the "Church" is synonymous with the pope and/or the bishops. "Magisterium" refers to the content of the teaching of those with an official obligation to teach in the Church. The pope, bishops, priests, and deacons all share this role by virtue of their office to preach, but usually the word is applied to the official pronouncements of pope and bishops.

5 As Carmen Barroso pointed out in her discussion of this paper in February 1990, debate about sex in Christianity is about power. Such debate has implications for gender roles and the maintenance of patriarchy. In addition, sex has been looked upon as sinful and morally dangerous. (See for example Pagels, 1988, on the development of sexual attitudes in Christianity.) The Church authorities take unto themselves a monopoly on forgiveness of sin. The ethics of sexual activity have a relationship, therefore, to the power of the clergy, who alone may give sacramental absolution from sin in Catholic practice.

6 Combining Western industrialized countries for the sake of analysis can result in disregard of important differences in religious behavior and beliefs. The United States, for example, differs greatly from France, Holland, and Germany with respect to religious belief and practice. Greeley (1990) examines some data concerning this variety.

7 The classic case, the Chinese rites controversy, pitted seventeenth-century missionaries who permitted veneration of ancestors in the Catholic liturgy against the Vatican Curia. Rome banned these practices as ancestor worship and idolatry. Proponents claimed they were a Chinese tradition not unlike the veneration of saints.

8 An old joke in the Vatican Curia asserts that in the Vatican there are no secrets, only Mysteries.

9 Personal communication from the commission member, 1990.

10 For an illustration, see Ratzinger, 1985, which is presented as a prolonged interview with a journalist, but is clearly contrived to allow the Cardinal to propose an authoritarian view of the Church.

## References

Barroso, Carmen, and Sonia Correa. 1990. "Politics of contraceptive research and testing in Brazil," paper presented at the conference on the Politics of Induced Fertility Change in Developing Countries, Bellagio, Italy, 19–23 February.

Carlen, Claudia. 1981a. *The Papal Encyclicals, 1740–1878.* Wilmington, NC: McGrath Publishing.

_____. 1981b. *The Papal Encyclicals, 1903–1939.* Wilmington, NC: McGrath Publishing.

_____. 1981c. *The Papal Encyclicals, 1958–1981.* Wilmington, NC: McGrath Publishing.

Day, Lincoln H. 1967. "Catholic teaching and Catholic fertility," in *Proceedings of the 1965 World Population Conference,* Paper 202. New York: United Nations.

Deen, Thalif. 1992. "UNICEF supports birth control despite Holy See," *Popline* 14, July–August: 3–4.

Donaldson, Peter J. 1990. *Nature Against Us.* Chapel Hill and London: University of North Carolina Press.

Finkle, Jason L., and Barbara B. Crane. 1975. "The politics of Bucharest: Population, development, and the new international economic order," *Population and Development Review* 1, 1: 87–114.

Greeley, Andrew. 1990. *The Catholic Myth.* New York: Macmillan.

Hehir, J. Brian. 1974. "The Church and the population year: Notes on a strategy," *Theological Studies* 35, 1 (March): 71–82.

Hoyt, Robert G. (ed.). 1968. *The Birth Control Debate.* Kansas City: The National Catholic Reporter Publishing Company.

Kaiser, Robert Blair. 1985. *The Politics of Sex and Religion.* Kansas City: Leaven Press.

Kelly, J.N.D. 1986. *The Oxford Dictionary of Popes.* New York: Oxford University Press.

Keohane, Robert O., and Joseph S. Nye, Jr. (eds.). 1972. *Transnational Relations and World Politics.* Cambridge, MA: Harvard University Press.

McBrien, Richard P. 1990. "A papal attack on Vatican II," *The New York Times,* 12 March, Op-ed.

McCormick, Richard. 1974. "The silence since *Humanae Vitae,*" *Linacre Quarterly* 41 (February): 26–32.

Miró, Carmen. 1963. "Características demográficas de America Latina," *CELADE,* Series A.E./C.N., A.12, D.3. 4/1. 4 Rev. 1.

Notestein, Frank W. 1971. "Reminiscences: The role of the Rockefeller Foundation, the Population Association of America, Princeton University and the United Nations in fostering American interest in population problems," *Milbank Memorial Fund Quarterly* 49, 4, part 2: 67–84.

Pagels, Elaine. 1988. *Adam, Eve, and the Serpent.* New York: Random House.

Piotrow, Phyllis T. 1973. *World Population Crisis: The United States Response.* New York: Praeger.

Potts, Malcolm, and Peter Selman. 1979. *Society and Fertility.* Estover, Plymouth, England: Macdonald and Evans.

Ratzinger, Joseph, with Vittorio Messori. 1985. *The Ratzinger Report: An Exclusive Interview on the State of the Church.* San Francisco: Ignatius Press.

Reese, Thomas. 1989. *Episcopal Conferences: Historical, Canonical and Theological Studies.* Washington, DC: Georgetown University Press.

Shannon, William H. 1970. *The Lively Debate.* New York: Sheed and Ward.

Steinfels, Peter. 1990. "Has Catholicism lost a chance to be our moral clearinghouse?" *The New York Times,* 8 April, Section 4, p.1.

Stoll, David. 1990. *Is Latin America Turning Protestant?: The Politics of Evangelical Growth.* Berkeley: University of California Press.

Stycos, J. Mayone. 1968. *Human Fertility in Latin America.* Ithaca: Cornell University Press.

_____. 1971. *Ideology, Faith, and Family Planning in Latin America.* New York: McGraw-Hill.

Symonds, Richard, and Michael Carder. 1973. *The United Nations and the Population Question, 1945–1970.* New York: McGraw-Hill.

Vallier, Ivan. 1972. "The Roman Catholic Church: A transnational actor," in Keohane and Nye (1972), pp. 129–152.

Warwick, Donald P. 1982. *Bitter Pills.* New York: Cambridge University Press.

# The Transnational Politics of Abortion

BARBARA B. CRANE

IN MOST DEVELOPING COUNTRIES, abortion has only recently emerged as a public policy issue. Even in countries that have had active population and family planning programs, the influence of traditional pronatalist culture and inherited colonial laws has limited the frequency of induced abortion and has discouraged organized efforts to expand the availability of services. In recent decades, however, changing social conditions have increasingly challenged traditional norms. Urbanization, increases in female education and labor-force participation, and inadequate contraceptive services in many developing countries have resulted in more unwanted pregnancies than ever before and have led increased numbers of women to seek to terminate their pregnancies, often at severe risk to their lives and health.[1]

Changes have also occurred in the political and legal environment. Argentina, Brazil, Chile, Mexico, Kenya, Nigeria, Zambia, Indonesia, the Philippines, and Turkey are among the developing countries where de jure or de facto abortion policies have become the subjects of new or renewed conflict in recent years.[2] Not all of these conflicts have resulted in legal changes, although some 20 developing countries have revised their abortion laws and regulations since 1980, most of them in the direction of expanding the indications for abortion (Cook, 1989; Henshaw, 1990).

In the past, the politics of abortion in most developing countries tended to conform to a pattern once characterized by an American political analyst as "the mobilization of bias." As he put it, "Some issues are organized into politics while others are organized out" (Schattschneider, 1960: 71). In effect, when powerful institutions do not consider a change in existing policy (or nonpolicy) as being in their interest, the issue in question is unlikely to be addressed. This situation is still true for abortion-related questions in some countries and in some decisionmaking arenas. Yet, as overall political competition increases in many countries, and as previously excluded groups gain access to the mass media and assume a greater political role, abortion issues are becoming more visible than they were in the past. The pressures for change in official policies or administrative practices come from multiple sources, including doctors concerned with mortality and morbidity from septic abortions; family planning professionals who believe abortion should be available both for health reasons and as a backup to contraceptive failure; women seeking to increase access to abortion as a reproductive right; and religious groups seeking to maintain or impose limits on what they regard as a criminal act.

In addition to these domestic pressures, as in other areas of public policy in developing countries, decisionmaking concerning abortion is strongly affected by external influences. Some of these influences are the result of the natural diffusion of ideas through education and, more immediately, through mass media reports on abortion-policy developments in industrialized countries. But more and more, the activities of organized groups based in the West, as well as the official foreign aid policies of the governments of the United States and European countries, are also contributing to the politicization of abortion in the developing world. Until the Clinton administration took office in January 1993, the US government officially opposed abortion "as a method of family planning," and directed its funds for overseas private family planning agencies only to those not involved in performing or "promoting" abortion (Blane and Friedman, 1990: 2). Other donor governments and agencies, in contrast, have supported abortion-related research, training, and technology.[3] At the same time, nongovernmental actors—associations affiliated with religions, family planning agencies, women's rights groups, and medical organizations—with vastly divergent perspectives are organizing actively with their counterparts in developing countries to influence relevant global deliberations and national policies.[4]

The purpose of this chapter is to examine more closely the changing global political environment of abortion. What is the history of global deliberations on abortion issues? What has been the impact of US government policy? Who are the major actors currently seeking to influence national and international policies, and what are their objectives? What are their strengths and weaknesses? The chapter concludes with an overview of the conditions that are likely to affect the future prospects for expanding access to abortion.

## The abortion issue in global perspective, 1960–92

Analysts often disagree over the importance to give to the declarations on social policy issues emanating from intergovernmental meetings. At a minimum, these pronouncements may serve as barometers of government attitudes; more significantly, they often help to establish a normative framework for policy development and provide guidance for the work of international agencies (Jacobson, 1984). These functions have been apparent in the many resolutions and declarations over the years in the area of family planning. As recently as the early 1960s, the subject of family planning evoked passionate controversy and was generally considered an inappropriate realm for government involvement. Eventually, governments found ways to address the subject in international forums and to define areas of agreement that enabled international agencies to undertake new programs (Symonds and Carder, 1973). With respect to abortion-related policies and programs, however, this consensus-building process has scarcely begun.

From the 1960s onward, discussions of contraception, family planning, and maternal and child health at the global level might have provided a logical context in which to address abortion policy. Among developing countries, China, Cuba, India, South Korea, Tunisia, and Zambia liberalized their national laws during the 1960s and early 1970s to allow the practice of abortion for nonmedical indications (Cook, 1989). Yet even as late as the World Population Conference in Bucharest in 1974, governments preferred to avoid the issue. In United Nations forums, where each nation's vote counts equally, the number of governments willing to participate in consensus is often more important than the size or power of the countries represented. Thus, the only mention of abortion in the World Population Plan of Action adopted at Bucharest was as part of a mild injunction to governments to work for "the reduction of involuntary sterility, subfecundity, defective births, and illegal abortions" (United Nations, 1975: 169). Significantly, the first International Women's Conference, held the following year, provided the occasion for comparatively extensive discussion of abortion, but more discussion occurred in the concurrent nongovernmental meetings than in the official sessions. As one participant in these meetings noted, the World Plan of Action for women did incorporate a statement "in a very neutral context" linking illegal abortion and maternal mortality (Oakley, 1975).

By the time governments convened at the second global population conference held in Mexico City in 1984, the political influence of abortion opponents was felt more strongly than ever before. International family planning leaders concerned about abortion had succeeded in including a statement in the draft declaration calling on governments to address the health consequences of illegal abortion. But at the conference, the Vatican delegation, supported by delegates from the United States and some third world countries, gained approval for a statement urging governments "to take appropriate steps to help women avoid abortion, which in no case should be promoted as a method of family planning, and whenever possible, provide for the humane treatment and counselling of women who have had recourse to abortion. . . " (United Nations, 1984: 767). While the phrasing was a compromise, and while supporters of access to abortion portrayed the admonition against promoting abortion as relatively meaningless, opponents perceived that it was a step toward officially expressing a norm against abortion except in cases of rape, incest, or to save the woman's life.[5] In calling for humane treatment for women who have had abortions, delegates removed the adjective "illegal" from the original draft, which redirected the meaning of the statement to imply that legal abortions are as much a potential source of suffering as illegal abortions. Based primarily on the Mexico City recommendations, the Governing Council of the United Nations Population Fund (UNFPA) affirmed as policy in 1985 that the fund would not provide assistance for abortion services "as a method of family planning"[6] (UNFPA, 1990).

From 1985 to 1991, no significant change occurred in the way abortion was treated in statements issued by intergovernmental forums—or even by less formal gatherings involving intergovernmental organizations. An international conference on safe motherhood in 1987 issued a call to action that identified the problem of unsafe abortion but made no direct recommendation that safe abortion services be made more available (Dixon-Mueller, 1990). In 1990, an expert meeting on the theme "from abortion to contraception" was cosponsored by UNFPA, the World Health Organization (WHO), and the International Planned Parenthood Federation (IPPF) in Tbilisi, in the former Soviet Union. UNFPA discouraged dissemination of the proceedings and the final declaration of the meeting, which called for safe abortion services.[7]

The political climate had the effect of making not just UNFPA, but other international organizations involved in family planning such as WHO, the World Bank, and IPPF extremely cautious in their approach to abortion questions. WHO, because of its medical orientation, was protected from some of the pressure felt by other agencies and, through its Human Reproduction Research Programme, was able to support research during the 1980s on new methods of inducing early abortion (WHO, 1978 and 1990).[8] In 1989, WHO also issued a call for research proposals on "the determinants and consequences of induced abortion." Yet WHO leaders avoided speaking out on the question of access to abortion services.

The World Bank provided loans for constructing health facilities in which abortions could be performed, but largely refrained from addressing abortion policies in the wide-ranging "sector" reports in which it reviews national health and demographic situations.

Because IPPF is a private voluntary organization, its secretary-general, Dr. Halfdan Mahler, could confront antiabortion activism in public statements more freely than could officials of other international agencies. Yet IPPF's efforts to expand access to abortion were constrained until recently by opposition from some of its national affiliates and by the heavy reliance of the federation and its affiliates on donor-government funding.[9] The organization was led from within to adopt in late 1992 the explicit goal of increasing access to safe, legal abortion (IPPF, 1993). IPPF is now expected to lay important groundwork for changes in norms and policies internationally and within countries where its affiliates are active. Additional expert meetings sponsored by UN agencies during 1992 in preparation for the

1994 international population conference provided further impetus for reconsideration of abortion.

## Abortion and US international population policy

To understand the changes in the international environment on the abortion issue over the past two decades, analysis must begin with the single most influential actor—the government of the United States. Until recently the most forceful proponent of family planning internationally, the US government was not always opposed to assisting abortion services. In the late 1960s and early 1970s, officials in the Office of Population of the US Agency for International Development (USAID) believed that access to methods for terminating early pregnancies was an essential element of effective family planning programs (Warwick, 1982). USAID devoted substantial funds to research on prostaglandins and on other technologies for early abortion, often referred to as "menstrual regulation" (Warwick, 1980; Laufe, 1979). Until 1973, USAID supported dissemination of new technologies to healthcare providers.[10] After 1973, however, a legislative amendment adopted by Congress at the urging of Senator Jesse Helms (Republican, North Carolina) prohibited the agency from using government funds for abortion-related purposes other than research. When the Reagan administration took office in January 1981, funding of all research on abortion methods was discontinued, although USAID was permitted to "gather descriptive epidemiological data to assess the incidence, extent or adverse consequences of abortion" (USAID, 1982: 187). During the period from 1973 to 1984, moreover, USAID was able to assist organizations in the developing world who were involved in abortion-related activities with non-USAID funds.

Under the Reagan presidency, legal restrictions on the use of USAID funds were gradually supplanted by a policy of active opposition to abortion. Administration leaders felt that antiabortion groups had been a key constituency in the presidential election (Hershey, 1986). The administration's determination to use the government's leverage against abortion became apparent in the draft position prepared for the US delegation to the International Conference on Population, held in Mexico City in August 1984 (Finkle and Crane, 1985; Fox, 1986). At the conference, the delegation articulated what has become known as the Mexico City Policy, stating in part that "the United States does not consider abortion an acceptable element of family planning programs" and reiterating US commitment to voluntarism in family planning (United States, 1984: 578).

Subsequently, the policy was implemented through decisions to withdraw funds from the IPPF and UNFPA. IPPF was deemed ineligible for funding because some of its affiliates were involved in providing abortion services. US contributions to UNFPA were terminated on the grounds that the agency supported projects in China, in effect sanctioning coercive abortion in the Chinese family planning program (Crane and Finkle, 1989). In addition, new restrictions prevented government funds from going to private organizations overseas that provided abortion services. As a result, Family Planning International Assistance (FPIA), a major agency of US support for family planning overseas, was unable to renew its five-year agreement with USAID in 1990 after losing a lengthy battle in the courts.[11] The Bush administration reaffirmed the Mexico City Policy, and efforts in Congress to overturn the policy failed.

Apart from its direct effects, the Mexico City Policy had a chilling effect on research and discussion related to abortion (Camp, 1987). Organizations receiving US government funds were greatly inhibited in initiatives they might have taken in the area of abortion, and many even avoided activities that were permitted by the policy, such as documenting the incidence of abortion.[12] Although UNFPA did not receive US funds after 1984, the agency stood apart from abortion issues, apparently because it desired to see US funds restored.

The rescinding of the Mexico City Policy by the Clinton administration in January 1993 opens the possibility for major changes in US policy with regard to support for abortion services, although political and legal constraints remain. Moreover, however significant they may be, official US actions are only one element of a dynamic international environment. Private transnational actors based in developed countries (mainly in the United States) have played an increasingly important role in the area of abortion issues, as have individuals and groups in developing countries who participate in transnational networks and coalitions. Students of international politics have long recognized the importance in many issue-areas of such groups and the "contacts, coalitions, and interactions across state boundaries that are not controlled by the central foreign policy organs of governments" (Keohane and Nye, 1971: xi). Although the directives of central foreign policymakers may set the broad parameters of an official position, as in the case of the Mexico City Policy, national political leaders are seldom directly involved in the foreign policy aspects of abortion issues. Few governments, and even less the societies over which they have authority, exhibit a well-articulated or unified perspective on abortion. As with other population and family planning issues, the potential exists for significant transnational influences (Crane and Finkle, 1988).

## The role of transnational actors

Appreciation of the heterogeneity of the organized groups actively involved in abortion issues is critical to understanding the politics of abortion internationally. Groups differ greatly in the way they are structured, their links to other institutions, their membership, and their command over resources.

Groups exhibit a vast range in what they regard as appropriate responses to unintended pregnancies and in their strategies for achieving their objectives. Still, for purposes of this discussion, two broad categories can be identified: (1) antiabortion groups, namely those who seek to proscribe, or strictly limit, access to abortion—and often, to contraceptive methods as well; and (2) prochoice groups, namely, those who support greater access to safe abortion and who affirm, sometimes with limitations, women's right to choose abortion.

### Antiabortion groups

Organized opposition to abortion, with grassroots support, began to emerge as a significant movement in Western countries only after abortion policies were liberalized in the United Kingdom (1967), the United States (1970–73), and Europe (mid-1970s). In the United States, critical events included the decisions of the US Catholic bishops in 1973 to create the National Right to Life Committee and in 1975 to launch the Pastoral Plan for Pro-Life Activities to facilitate grassroots organizing (Paige, 1983). In other Western countries, religious leaders opposed to abortion confined themselves mostly to expressing their views without investing resources in political organizing on the issue. As public support around the world for population and family planning programs grew in the early 1970s, the Vatican was concerned with the threat the programs posed to Church authority (Keely, this volume). In the period before the Bucharest World Population Conference in 1974, the Vatican, therefore, strengthened its organizational units concerned with reproductive issues.[13] Still, these issues tended to be eclipsed by other concerns, and the Vatican was unprepared to press them in the face of both differences within the Church and the national population policies that had been adopted prior to 1974 (Keely, this volume).

The situation changed in the late 1970s and early 1980s, providing the impetus for a transnational antiabortion movement. One development was the selection in 1978 of a new Pope, John Paul II, who saw promotion of the Church's doctrine on reproductive issues, especially in the areas of abortion and strengthening of the family, as a high personal priority. He frequently expressed his position during his travels to developing countries, which were more extensive than those of any previous pontiff. More important, he was willing to reward bishops and priests who shared his zeal and he approved strong sanctions against those who dissented from Catholic doctrine.[14] In addition, the pope helped to strengthen Opus Dei, a transnational organization of Catholic clergy and lay people, whose members have been leading opponents of expanding access to contraception and abortion. The Vatican delegation to the Mexico City conference in 1984 worked successfully on behalf of antiabortion amendments and led opposition to policies such as those of China and India that, in its view, constituted "promotion of abortion as a method of family planning."[15] The Vatican also changed its posture during the 1980s from passive endorsement of "natural family planning" (periodic abstinence) to active encouragement of this practice for married couples desiring to limit fertility. In 1980, US Catholic leaders persuaded Senator Frank Church (Democrat, Idaho), chairman of the Senate Foreign Relations Committee, to sponsor an amendment mandating USAID to support relevant information and services. Adoption of the amendment led USAID to fund groups with antiabortion views to provide natural family planning, enabling them to strengthen their international networks (Benshoof, 1987). The decisions and actions of Church-affiliated organizations are important in a number of developing countries, not only because of their moral authority but also because they are autonomous from government control and often play a central role in the delivery of health services.

The second major development contributing to a transnational antiabortion movement was the growth of organized opposition to abortion in the United States, reaching beyond Catholics to attract some Protestant support as well, with the help of such prominent "New Right" conservative leaders as Senator Jesse Helms and Paul Weyrich (Paige, 1983). While US abortion opponents had only limited success in achieving their domestic objectives, they were able to exercise a disproportionate influence on official US decisionmaking in the vulnerable area of foreign aid policy and were also able to use their resources to reach out transnationally.[16] The 1984 Mexico City conference became a platform from which to gain global attention for antiabortion perspectives and an opportunity to strengthen transnational connections, especially with groups from other Western countries that were represented at the conference.

Subsequently, the International Right to Life Federation was founded in 1986, obtained consultative status with the United Nations, and opened offices in Lausanne and Rome; its president is also president of the US National Right to Life Committee. The most active such organization internationally, with local offices in a number of countries including Brazil, Nigeria, the Philippines, and Zambia, is Human Life International (HLI). This organization, headed by a Catholic priest, works closely both with local Catholic hierarchies and with the militant wing of the antiabortion movement in the United States.[17] HLI holds international conferences (for example, in Kenya in 1985, Mexico in 1987, Zambia in 1989, and Czechoslovakia in 1992) and it disseminates publications and films to promote its cause.

The impact of the antiabortion movement has been felt in the formation of public policies in the United States, the Philippines, and Brazil. It has also delayed decisions by German and French pharmaceutical companies to seek foreign governments' approval for marketing RU 486, a medication that makes early abortions possible without surgery.[18]

The activities of antiabortion groups around the world tend to be dominated by Catholics and are grounded in a highly coherent set of principles laid out in such statements as the papal encyclical of 1968, *Humanae Vitae*. These principles hold that "the direct interruption of the generative process already begun, and, above all, directly willed and procured abortion, even if for therapeutic reasons, are to be absolutely excluded as licit means of regulating births" (Pope Paul VI, 1968). Differences of perspective exist among Catholic thinkers over tactical questions of whether to support restrictive measures that stop short of a total ban on abortion.[19] Activist antiabortion groups have, nevertheless, tended to be dominated by those who oppose abortion under any circumstances and who uphold natural family planning by married couples as the only legitimate means to avoid pregnancy. Not until the debate over abortion becomes more advanced internationally and within developing countries are differences among abortion opponents likely to inhibit their effectiveness.

### Prochoice groups

Attention to family planning issues began to gather momentum internationally through the efforts of private groups in the early 1960s; active participation by governments and international agencies began in the middle to late 1960s. These agencies devoted little attention to the problem of access to abortion until the early 1970s, when both technological developments and movements to legalize abortion in the United States and other Western countries converged to present new opportunities. The most important technological development, because of its potential for application in developing countries, was the Karman cannula, a simple device for removing the contents of the uterus in early pregnancy (Lee and Paxman, 1975). Research on other abortion techniques was also yielding results. India had legalized abortion, and governments of two of the other largest developing countries, Bangladesh and Indonesia, had decided to allow menstrual regulation (Lee and Paxman, 1975). The active role in matters pertaining to abortion that was initially assumed by the USAID Office of Population had to be relinquished fairly early, however, in the interest of retaining congressional support for family planning.[20] Further international work on abortion would depend much more on initiatives from a range of other groups holding prochoice views—private population and family planning donor agencies, women's rights groups, and medical and public health organizations.

*Population and family planning donor agencies* Most of the major private organizations involved in international family planning have been reluctant to fill the gap left by USAID.[21] For those organizations with demographic objectives, efforts to expand the practice of contraception have seemed to hold far more promise as an approach to reducing fertility than efforts to promote the safety and legality of abortion. Leaders of these organizations have felt that abortion-related projects and programs invite confrontation and endanger their efforts to promote contraception. Funding for abortion-related activities is estimated to account for less than one percent of public and private donor spending on population and family planning assistance.[22]

A few population and family planning specialists have given higher priority than their colleagues to expanding access to safe abortion, especially in those countries where abortion is already legal or where legal restrictions are ambiguous or rarely enforced. The International Projects Assistance Service (IPAS), founded in 1973, and the International Women's Health Coalition (IWHC), founded in 1980, have been the nongovernmental agencies most involved in efforts to promote greater access to safe abortion in developing countries. The Population Crisis Committee (now Population Action International), a Washington-based organization, was instrumental in founding IWHC and helped channel funds into the work of both of these agencies. As official US policy moved toward an actively anti-abortion posture, both IPAS and IWHC expanded their programs with additional private foundation support. Another private organization founded in Britain in the 1970s, Marie Stopes International (formerly Population Services Europe), assists family planning clinics in some South Asian and African countries in the areas of abortion and contraception.[23]

Family planning specialists who support prochoice activities believe that safe abortion services, like contraception, contribute to maternal and child health and well-being and should be regarded as a normal component of maternal and child health and family planning programs. Family planners further contend that abortion will be necessary even when contraception has improved, because people exercise foresight less effectively than hindsight.[24] Moreover, contraceptives themselves still often fail.[25] Some specialists argue, therefore, that it is unethical to provide contraceptives in national family planning programs without also ensuring that abortion is available when contraceptives fail. Finally, many also believe, because of the historical experience of such countries as Cuba and Korea, that expanding access to both safe abortion and contraceptive services is likely to have a significant impact on fertility.[26]

Organizations that identify with these propositions in principle have been slow to follow through on their implications. In the first few years after the adoption of the restrictive US policy in 1984, hopes persisted that the policy could be overturned and that the political climate would change under the presidential administration taking office in 1988. The official US position remained unchanged, however, and private organizations such as the Pathfinder Fund, the Population Council, and others began taking initiatives with private funds, although they were conscious of potential repercussions on their contracts with the US government.

After RU 486 began to be distributed in 1988, family planning specialists had hopes that it would transform the policy debate and would lessen the need for expensive clinical services.[27] These hopes have diminished as it became apparent that close medical supervision is necessary for RU 486 to be used safely and effectively and that further research is required to determine whether the drug would provide a safer and more accessible option than vacuum aspiration for women in developing countries (WHO, 1990). In developing countries, promoting abortion-related research has been easier than finding support for expanding services. After losing its US government funding, the international arm of the Planned Parenthood Federation of America decided to increase its role in abortion services, but found that resources were difficult to obtain; private foundations are often reluctant to fund services in general, particularly those involving abortion.[28] While some support for international abortion services in developing countries can be obtained from private sources, significant expansion depends on funding from donor or developing country governments.

Even family planning agencies that are willing to conduct research and to provide training and services have been reluctant to attempt to influence national abortion laws and regulations. Agencies disagree about how much effort to devote to policy changes and what strategies to pursue. Some believe that emphasis should remain on expanding services within the boundaries of existing policies.[29] This view is consistent with the belief that focusing on services allows the agency to avoid direct confrontation with policymakers and with organized opposition while it paves the way for future policy change. Other agencies, such as IPAS, prefer to focus on administrative regulations and practices that can affect access to abortion; they work for incremental and de facto policy change without creating opportunity for divisive public debate.[30] Still others, including Catholics for a Free Choice (CFFC), avoid becoming involved in abortion services or administrative issues and seek instead to shape public discourse and thereby the broader policy environment of abortion.[31] Some prochoice groups in developing countries, including Brazil, also emphasize the need to change official attitudes and policies, believing that without public funding, removing restrictive laws or ending their enforcement would have limited impact on poor women's access to abortion.[32]

The ambivalence of family planning agencies concerning abortion and differences within and among them about how best to proceed have made them slow in forming a strong and visible coalition to promote access to abortion. The antagonisms persist, as well, between some family planning groups and their potential coalition partners: women's rights advocates and members of the medical community.

*Women's rights advocates* Women's networks and organizations only began to become an important political presence internationally during the 1970s. The emergence of transnational leadership networks dedicated to advancing women's role in development through new programs fostered by public and private donors contributed to this trend. These networks emphasized women's education and labor force participation, however, paying comparatively little attention to reproductive issues (Ashworth, 1982; Boserup, 1970).

The first international meetings on women's health were held in the late 1970s in response to the growing women's health movement in the United States and other Western countries during the previous decade. These meetings provided a natural forum in which reproductive issues could be addressed. Women were mobilized by their perception that family planning programs were often coercive; feminist observers criticized programs for their inattention to the consequences of many modern contraceptives to clients' health and safety (Hartmann, 1987; Petchesky, 1984; Gordon, 1990). These concerns led to the founding of the International Contraception, Abortion, and Sterilization Campaign (ICASC) in 1978, initially based in London; in 1984, it became the Women's Global Network for Reproductive Rights (WGNRR), based in Amsterdam. Drawing support from private foundations, WGNRR and related groups have issued publications and have helped to organize influential international meetings, in Geneva (1981), Amsterdam (1984), Costa Rica (1987), and Manila (1990) (WGNRR, 1985–92).[33]

In developing countries, women's health groups began to form in the 1980s, gaining particular strength in Latin America. The United Nations International Women's Conference in Nairobi in 1985 was a significant event, bringing together not only official delegates but also representatives from feminist groups around the world.[34] Since then, the International Women's Health Coalition, the Women's Global Network for Reproductive Rights, and the Boston Women's Health Book Collective have been assisting national and regional groups in developing countries in organizing international meetings, disseminating information, and creating documentation centers.

Women's rights groups differ in their willingness to work with family planning agencies. In recent years, however, spurred by the need to counter antiabortion groups, efforts have been made to bridge differences and agree on a comprehensive approach to women's reproductive health issues. Adherents of this approach believe that starting with the needs of individual women will also contribute to reducing fertility (Dixon-Mueller, 1993; Germain, 1987; Germain and Ordway, 1989). The reproductive health approach includes maternal and child health care, treatment for diseases and infections of the reproductive tract, sex and health education, contraceptive services for women regardless of age or marital status, and abortion services. Emphasis on this approach is often believed to provide a justification that will be more widely acceptable politically for expanding abortion access. To the extent that this approach ties abortion services to a particular model of women's reproductive health care, however, and skirts the question of what legal restrictions may be placed on abortion, it may not satisfy those who feel access to safe abortion needs to be established in the public mind as a woman's right.

Regardless of what approach they take, both local and transnational women's groups are relatively weak. The women whose reproductive health needs are greatest are also those most difficult to organize politically and least likely to be able to lead and sustain effective efforts to influence policy. The impact of women's health groups in supporting access to abortion is, therefore, most likely to be felt if they can combine their efforts with those of other organized prochoice groups. The challenge for women's groups concerned with reproductive rights and health, however, is to find allies who are willing to be as outspoken as they are.

*The medical community* Most physicians agree in principle, some with reservations, that women should have access to safe abortion services.[35] Medical and public health associations in the United States in the 1960s, for example, took official positions that were a critical factor in the movement for abortion reform even before the 1973 Supreme Court decision (Luker, 1984). Special initiatives have been undertaken in recent years to engage the international medical community in translating their prochoice principles into reality. Since 1985, family planning specialists have collaborated with health professionals in such international organizations as WHO, the World Bank, and UNICEF in new efforts to reduce maternal mortality in developing countries. While "safe motherhood" is seen as an important objective in its own right, these efforts have also served to bring more attention to the role of septic abortion as a health issue (Rosenfield et al., 1989). Other initiatives have been pursued through medical networks and associations. For example, triennial meetings of the International Federation of Gynecologists and Obstetricians (FIGO) in Berlin (1985) and Rio de Janeiro (1988) were each followed by a special conference on women's reproductive health, including abortion, organized by the International Women's Health Coalition (Rosenfield et al., 1989). The appearance of RU 486 as a new means of inducing early abortion has helped to mobilize physicians politically because they perceive a threat to scientific and medical progress associated with antiabortion influence over dissemination of a drug that is expected to have a variety of medical uses. In 1988, a petition signed by more than 1,000 FIGO members gathered in Brazil convinced the French government to reverse its earlier decision and approve production and distribution of RU 486 in France.[36]

Despite these recent developments, mobilizing the medical profession to push for expanded access to abortion has proved to be a slow process. Many physicians who are trained to perform abortions are reluctant to do so and have low regard for practitioners who provide services.[37] The skills involved in providing abortion care are not those that command the highest respect within the medical profession (Davis, 1985). Moreover, because of legal restrictions and social attitudes toward abortion, the service remains clandestine or partially so in many countries. As a result, the providers are often marginal practitioners motivated by profit or are so perceived by physicians, by the general public, and even by some prochoice activists.[38] Although providers of clandestine abortions might be expected to support removal of legal restrictions, in fact they might hesitate to promote changes that could threaten the services they are providing, or, in some cases, their profits (Leonard, 1989). Physicians who have profited by providing surgical abortions may resist learning less expensive menstrual regulation procedures and may oppose measures to enable nonphysicians to provide abortions.[39] Physicians who might be interested in providing abortions are often deterred because they are not schooled in the interpretation of relevant laws and regulations (Germain, 1989).

Legitimate abortion providers have no visible international presence and tend to be minor actors in abortion policymaking even in the United States, where most are members of the National Abortion Federation (NAF).[40] Although it exists to improve the quality of abortion care and facilitate the flow of relevant technical information, NAF is often perceived as a "trade union" for abortion providers. Only in the last few years has the organization begun to play a more effective role in abortion policy debates. NAF has had some dealings with abortion providers in other countries, especially in Canada, but has been concerned almost entirely with abortion service delivery in the United States. In other countries where abortion is legal, procedures are mainly performed in hospitals rather than in outpatient clinics. Therefore, the potential for new national or transnational associations of providers is uncertain. Whether the existing medical associations or networks with members from developing countries will become a significant force for expanding access to abortion remains to be seen.

Clearly, transnational groups identified with prochoice views differ considerably in their structures and priorities. To some degree, this pluralism is an advantage, because each group can draw on different political resources to advance the cause of expanded access to abortion. But when prochoice groups are openly critical of each other's motives and actions, they undermine one another. Differences are exacerbated by the overall scarcity of resources for international family planning and women's health programs.[41] Obstacles to cooperation tend to be even greater within developing countries. Groups begin with different priorities and limited information; they have difficulty defining a common approach on abortion in a setting where resources are extremely scarce and where the costs of being outspoken on a controversial issue may be much higher than in a developed country or in an international forum.

## Prospects for future access to abortion

The international environment for abortion issues is fluid at present. Both prochoice and antiabortion transnational groups may find that they need to modify their strong policy positions in order to gain resources and find allies among groups in developing countries where perspectives on abortion policy are still evolving. The future prospects for access to abortion are comparable to those of international family planning activities 30 years ago. In 1963, a few governments were supportive of family planning, and a number of private organizations were active in providing services; yet initiatives were tentative and small in scale. An environment hostile to family planning was perceptible in most countries (Symonds and Carder, 1973).

What changes in the environment of abortion policy are likely to affect the prospects for expanding access to abortion in developing countries? First, although the impact of a US turnaround on abortion-related policies may be less now than such a turnaround was on contraception-related policies at the height of US global influence two or three decades ago, it is still likely to make a difference. Since early 1993, the Clinton administration has not only rescinded the Mexico City Policy, but has openly stated its commitment to reproductive choice, including access to safe abortion. The Helms Amendment remains in effect, however, prohibiting the use of foreign-assistance funds to pay for abortions "as a method of family planning."

Changes in international priorities could also foster a more supportive policy climate. The United Nations organizers of the International Conference on Population and Development (ICPD) in 1994 expect that abortion will figure prominently on the agenda. With support from most donor governments, the ICPD Secretary-General, Nafis Sadik (also Executive Director of UNFPA), has called for recommendations aimed at eliminating illegal and unsafe abortion. While the ICPD cannot impose a particular policy on any government, it could endorse prochoice policies and give international agencies a wider mandate to respond to government requests concerning research, training, and technical assistance for safe abortion services and counseling. Potential opposition to these policy changes may be significant for a time, however, depending on the extent to which governments base their positions on Catholic or other religious doctrines that severely restrict or prohibit abortion.

For the long term, other forces that will likely contribute to the changing environment, as they did for family planning generally in the 1960s, include the greater availability of improved technologies and wider social acceptance of abortion associated with changes in women's roles and in cultural values.[42] Abortion policy will vary according to country, of course, and each will respond in terms of its own culture, religion, and politics, but few countries will remain unaffected by trends in the West.

Both antiabortion and prochoice groups will find opportunities to advance their causes. The major question for groups opposed to abortion is whether they will seek to broaden their appeal by supporting measures to limit but not to prohibit abortion. A willingness to compromise may increase their immediate influence but will also undermine their longer-term goal of eliminating abortion. This issue is difficult to resolve because the political force of these groups is their unified commitment to a consistent moral position, their fervor in support of their cause, and their credibility as morally motivated individuals who will not surrender to expediency. As they become involved in national debates, it is particularly important for antiabortion groups with transnational connections to preserve the legitimacy they gain by claiming to support universal moral precepts.

For prochoice groups, the challenges are different. In order to increase access to safe abortion, these groups must agree on the procedures by which women with unwanted pregnancies should make and implement decisions to terminate them and on the role of public policies and programs in these decisions.[43] Because medical intervention is generally necessary to carry out abortion safely, expanding access to safe abortion also entails more than removing legal restrictions or changing public attitudes; resources are needed to support high-quality services as well as the individuals and institutions willing and able to provide them. The capacity of prochoice groups to advance their cause will depend on whether they reconcile or put aside their differences with one another and commit substantial political and financial resources to their common objectives. If these steps are taken, access to safe abortion is likely to expand as rapidly in the decades ahead as access to modern contraception did in the decades past.

## Notes

The author thanks the Ford Foundation and the Harvard School of Public Health for generously supporting most of the research on which this article is based. The assistance of those individuals involved in abortion issues who participated in interviews and shared relevant documents is especially appreciated. Much of this article relies on the understanding they have given the author, even where no specific citations are provided. Confidentiality of some individuals cited concerning this sensitive subject has been respected by using their organizational affiliations rather than their names. Finally, the author is most grateful for the helpful comments received on earlier versions of the article from Lincoln Chen, Henry David, Ruth Dixon-Mueller, Adrienne Germain, Charles Keely, Frances Kissling, Michael Reich, Deborah Rogow, and Joseph Speidel.

1 Induced abortion has been practiced throughout history, even in cultures where prevailing norms have opposed it. During the demographic transition to lower fertility, population specialists suggest that either abortion alone, or both abortion and contraception, tend to increase for a time until eventually "contraceptive practice improves, and resort to abortion declines, although it is never eliminated" (David, 1983: 208; Requena, 1970). Current estimates of abortion incidence in developing countries are based on limited data, making it difficult to determine the extent to which contraceptive availability is helping to hold down the incidence of abortion. The annual number of deaths attributable to clandestine abortions in developing countries is believed to be close to 100,000, with World Health Organization estimates being as high as 200,000, or between 20 and 40 percent of all maternal deaths (Henshaw, 1990). The extent to which illegal practitioners are adopting safer methods is unknown. Research is being undertaken to confirm the impression that in some countries the situation with regard to septic abortions is improving slowly or may be worsening (Dixon-Mueller, 1990; Rosenfield et al., 1989; Figa-Talamanca et al., 1986; Coeytaux, 1988).

2 Confirmation of this point comes from the author's interviews with knowledgeable observers, unpublished trip reports and professional conference presentations, and such periodical publications as *Abortion Research Notes* (David, 1984–92) and the newsletters of the Women's Global Network for Reproductive Rights (1985–92). For further background on Brazil, Nigeria, and the Philippines, see the chapter by Dixon-Mueller and Germain in this volume.

3 Reports from informed observers in interviews with the author suggest, however, that the only bilateral donors involved in cautiously supporting abortion-related activities are the governments of Finland, Great Britain, the Netherlands, and Sweden. No other donor followed the US lead in withholding funds from organizations involved in abortion with private resources, although Canada and Germany have specified that their contributions to the International Planned Parenthood Federation cannot be used for abortion activities.

4 In some Eastern European countries, abortion policy issues have recently been reopened under new governments. External actors, including international agencies and private groups, are also becoming more involved (David, 1984–92). As a result, although this article focuses on the developing countries of Asia, Africa, and Latin America, much of the analysis applies to Eastern European countries as well.

5 See Finkle and Crane, 1985. In a 1974 policy statement USAID had precluded support for "programs that seek to promote abortion as a method of family planning" (Warwick, 1980: 305).

6 During the 1970s, UNFPA had provided abortion-related assistance to at least one developing country, Tunisia, where abortion is legal (Warwick, 1982).

7 See the report of the Tbilisi meeting (WHO, 1991). The Tbilisi conference stands in contrast to the donor-sponsored International Conference on Family Planning in the 1980's, held in Jakarta a decade ago, when more than 130 invited participants from 63 countries issued a statement that included the following recommendation: "Since most currently available contraceptive methods fall short of providing complete protection against unwanted pregnancy, access to safe, modern abortion techniques should also be offered wherever laws permit" (UNFPA et al., 1981: 35).

8 During the 1970s, WHO convened expert groups on abortion and provided some assistance for abortion training (WHO, 1978).

9 Interview with Allan Rosenfield, Dean, Columbia University School of Public Health, New York, November 1989.

10 Interview with Joseph Speidel, former USAID official and president of the Population Crisis Committee, Washington, DC, April 1991.

11 *The New York Times*, 9 March 1990. FPIA is administered by the Planned Parenthood Federation of America.

12 A study commissioned by USAID of projects it supported in six countries found that most were not significantly affected by the policy; but in some of them, "project management have reacted to Mexico City policy requirements by approaching the abortion

question with an overcautiousness that extends to activities clearly permitted under the policy" (Blane and Friedman, 1990: vi). See also Camp, 1987.

13 Interview with Monsignor James McHugh, former Director of the Office of Pro-Life Activities, US National Conference of Catholic Bishops, Newark, New Jersey, August 1987.

14 See *The Boston Globe*, 3 December 1989; *The New York Times*, 8 November 1989; and Keely, this volume.

15 Interview with Monsignor James McHugh, August 1987.

16 For an early attack on the role of abortion in US foreign aid, see a statement by the head of Americans United for Life originally delivered at the Congress for the Family of the Americas in Guatemala in 1980 (Trueman, 1980). The Congress, organized by the head of a private agency that would eventually receive USAID funds, included among its participants Mother Teresa; John and Evelyn Billings, inventors of the leading natural family planning method; Colin Clark, the development economist; and a personal representative of Pope John Paul II.

17 Interviews with staff members of Human Life International and the Population Crisis Committee; newsletters and brochures of Human Life International. One HLI newsletter notes that "there is a Pro-Life Association of Nigeria (PLAN), which the Catholic Bishops Conference in Nigeria actively supports," and then expresses gratitude to a Father Marx for helping to launch a "Human Life Protection League" (*HLI Reports*, 8 June, 1990: 3).

18 *The New York Times*, 26 March 1989.

19 *The New York Times*, 12 August 1989. Catholic leaders opposed to abortion also differ on questions of population policy and contraception; some, including many members of local hierarchies in developing countries, accept the legitimacy of voluntary programs that seek to make effective methods of contraception available. For examples of writing from this perspective, see Warwick, 1982 and Hehir, 1974, cited in Keely, this volume. A number of Catholic leaders and theologians go even further in endorsing the morality of early abortion under limited circumstances, and some treat abortion as a personal moral decision that may be justified under many circumstances (Maguire, 1986).

20 Interview with Jack Sullivan, former staff member of the US House of Representatives Committee on Foreign Affairs, Washington, DC, July 1985. USAID did sponsor a conference on abortion in 1978 attended by nearly 200 medical and family planning experts, including representatives from more than a dozen developing countries. See Zatuchni et al., 1979.

21 The former director of USAID's Office of Population has stated that in 1974 major foundations recoiled from his proposal that they support dissemination of menstrual regulation technology. Reimert T. Ravenholt, remarks at USAID Cooperating Agency meeting, Washington, DC, November 1990. See also Warwick, 1982.

22 Tomas Frejka, the Population Council, March 1991, personal communication.

23 Interview with Timothy Black, Population Services Europe, London, May 1986.

24 Interviews with staff members of the Population Crisis Committee and former officials of the USAID Office of Population, Washington, DC, 1990.

25 Currently, in the United States, rates of contraceptive failure (accidental pregnancy in the first year of use) range between 3 and 23 percent, according to method, for pills, IUDs, condoms, diaphragms, sponges, periodic abstinence, and vaginal contraceptives. These rates are even higher for developing countries (Mastroianni et al., 1990).

26 The impact on fertility of expanding access to safe abortion depends on how many births thereby averted would have been averted by other means—contraception, abstinence, or resort to unsafe (and often illegal) abortions—in the absence of greater access (Frejka, 1985). The impact would depend also on whether improved abortion services are linked with postabortion contraceptive counseling services. The experiences of the Soviet Union, Eastern Europe, and Japan, all areas where abortion has been demographically significant, were associated with limited access to effective means of contraception,

and, according to family planning specialists, do not provide a desirable model for developing countries. Turkey, which in 1983 adopted a new law to make abortion available on request in the first trimester, may provide a further test of what happens when contraception and abortion services are expanded simultaneously (Leonard, 1989).

27 Interview with Marie Bass, Reproductive Health Technology Project, Washington, DC, December 1990.

28 Interviews with staff members of Family Planning International Assistance, January and November 1990.

29 Interview with Jeannie Rosoff, President, Alan Guttmacher Institute, Washington, DC, January 1991.

30 Interview with Judith Senderowitz, IPAS board member, Washington, DC, February 1991.

31 Interview with Frances Kissling, President, Catholics for a Free Choice, Washington, DC, December 1990. CFFC, which is based in Washington, has been active in Latin America and has an office in Uruguay.

32 Interview with Carmen Barroso, former President, Committee on Reproductive Rights, Brazilian Ministry of Health, Boston, November 1989.

33 A few women's groups opposed to abortion exist, although in general antiabortion women advance their cause as members of organizations led primarily by men.

34 Interview with Norma Swenson, Boston Women's Health Book Collective, Boston, December 1989. A total of 167 women from 43 countries attending concurrent nongovernmental sessions signed a declaration calling for access to family planning, including abortion.

35 Some national and international organizations of antiabortion physicians exist, but they tend to be overshadowed by other groups.

36 *The New York Times*, 29 October 1988.

37 According to one observer, "The abortion clinic in American society sets itself apart from obstetrical delivery and care, and OB/GYN's have found numerous ways to keep abortion practice isolated from obstetric practice" (Imber, 1986: 9).

38 Interviews with Barbara Radford, Executive Director, National Abortion Federation, Washington, DC, December 1990 and Warren Hern, Director, Boulder Abortion Clinic, Washington, DC, March 1991.

39 Interview with Allan Rosenfield, Dean, Columbia University School of Public Health, New York, November 1989.

40 Interview with Barbara Radford, Executive Director, National Abortion Federation, Washington, DC, December 1990.

41 Interview with Jeannie Rosoff, President, Alan Guttmacher Institute, Washington, DC, January 1991.

42 In comparative studies of Western countries, survey researchers have found that the rise of a distinct set of "postmaterialist" values is associated with much more permissive attitudes concerning abortion (Inglehart, 1990).

43 A thoughtful discussion of these issues is contained in a presentation to the 1990 annual meeting of the National Abortion Federation by Frances Kissling, President, Catholics for a Free Choice (Kissling, 1990).

## References

Ashworth, Georgina. 1982. "The United Nations Women's Conference and international linkages in the women's movement," in *Pressure Groups in the Global System: The Transnational Relations of Issue-Oriented Non-Governmental Organizations*, ed. Peter Willetts. London: Frances Pinter.

Benshoof, Janet. 1987. "The establishment clause and government-funded natural family planning programs: Is the constitution dancing to a new rhythm?" *Journal of International Law and Politics* 20, 1: 1–33.

Blane, John, and Matthew Friedman. 1990. "Mexico City Policy Implementation Study," *Population Technical Assistance Project Occasional Paper* No. 5, Washington, DC.

Boserup, Ester. 1970. *Woman's Role in Economic Development*. London: Allen and Unwin.

*The Boston Globe*. 3 December 1989.

Camp, Sharon. 1987. "The impact of the Mexico City policy on women and health care in developing countries," *Journal of International Law and Politics* 20, 1: 35–52.

Coeytaux, Francine M. 1988. "Induced abortion in sub-Saharan Africa: What we do and do not know," *Studies in Family Planning* 19, 3: 186–190.

Cook, Rebecca J. 1989. "Abortion laws and policies: Challenges and opportunities," *International Journal of Gynaecology and Obstetrics* Supplement No. 3: 61–88.

Crane, Barbara B., and Jason L. Finkle. 1988. "Population policy and world politics," paper presented to the International Political Science Association, Washington, DC, September.

———, and Jason L. Finkle. 1989. "The United States, China, and the United Nations Population Fund: Dynamics of US policymaking," *Population and Development Review* 15, 1: 23–59.

David, Henry P. 1983. "Abortion: Its prevalence, correlates, and costs," in *Determinants of Fertility in Developing Countries*, eds. R. A. Bulatao and R. D. Lee, vol. 2. New York: Academic Press, pp. 193–244.

——— (ed.). 1984–92. *Abortion Research Notes*. Bethesda, MD: Transnational Family Research Institute.

Davis, Nanette J. 1985. *From Crime to Choice: The Transformation of Abortion in America*. Westport, CT: Greenwood Press.

Dixon-Mueller, Ruth. 1990. "Abortion policy and women's health in developing countries," *International Journal of Health Services* 20: 297–314.

———. 1993. *Population Policy and Women's Rights: Transforming Reproductive Choice*. Westport, CT: Praeger.

Figa-Talamanca, Irene, et al. 1986. "Illegal abortion: An attempt to assess its cost to the health services and its incidence in the community," *International Journal of Health Services* 16, 3: 375–389.

Finkle, Jason L., and Barbara B. Crane. 1985. "Ideology and politics at Mexico City: The United States at the 1984 International Conference on Population," *Population and Development Review* 11, 1: 1–28.

Fox, Gregory H. 1986. "American population policy abroad: The Mexico City abortion funding restrictions," *Journal of International Law and Politics* 18, 2: 609–662.

Frejka, Tomas. 1985. "Induced abortion and fertility," *Family Planning Perspectives* 17, 5: 230–233.

Germain, Adrienne. 1987. "Reproductive health and dignity: Choices by Third World women," paper prepared for the International Conference on Better Health for Women and Children through Family Planning, Nairobi, Kenya. New York: The Population Council.

———. 1989. "The Christopher Tietze International Symposium: An overview," in *International Journal of Gynaecology and Obstetrics* Supplement 3: 1–8.

———, and Jane Ordway. 1989. *Population Control and Women's Health: Balancing the Scales*. New York: International Women's Health Coalition and the Overseas Development Council.

Gordon, Linda. 1990. *Woman's Body, Woman's Right: Birth Control in America*. Revised edition. New York: Penguin Books.

Hartmann, Betsy. 1987. *Reproductive Rights and Wrongs: The Global Politics of Population Control and Contraceptive Choice*. New York: Harper and Row.

Henshaw, Stanley. 1990. "Induced abortion: A world review, 1990," *International Family Planning Perspectives* 16, 2: 59–65.

Hershey, Marjorie Randon. 1986. "Direct action and the abortion issue: The political participation of single-issue groups," in *Interest Group Politics*, eds. Allen J. Cigler and Burdett A. Loomis. Washington, DC: Congressional Quarterly.

HLI Reports. 1984–89. *HLI Reports*. Rockville, MD.: Human Life International.

Imber, Jonathan B. 1986. *Abortion and the Private Practice of Medicine*. New Haven: Yale University Press.

Inglehart, Ronald. 1990. *Culture Shift in Advanced Industrial Society*. Princeton: Princeton University Press.

International Planned Parenthood Federation (IPPF). 1993. *Planned Parenthood Challenges: Unsafe Abortion*. London: IPPF.

Jacobson, Harold K. 1984. *Networks of Interdependence: International Organizations and the Global Political System*. New York: Alfred A. Knopf.

Keohane, Robert O., and Joseph S. Nye, Jr. (eds.). 1971. *Transnational Relations and World Politics*. Cambridge, MA: Harvard University Press.

Kissling, Frances. 1990. "Ending the abortion war: A modest proposal," in *Reaching Out: Speaking for Abortion in the '90s*. Washington, DC: National Abortion Federation.

Laufe, Leonard. 1979. "Menstrual regulation—international perspectives," in Zatuchni et al., 1979.

Lee, Luke T., and John M. Paxman. 1975. "Legal aspects of menstrual regulation: Some preliminary observations," *Journal of Family Law* 14, 2: 181–221.

Leonard, Ann, et al. 1989. "Turkey's liberalized abortion law: A policy response to illicit abortion practices," paper presented to the Annual Meeting of the American Public Health Association, Chicago, Illinois, October.

Luker, Kristin. 1984. *Abortion and the Politics of Motherhood*. Berkeley: University of California Press.

Maguire, Marjorie R. 1986. "Pluralism on abortion in the theological community: The controversy continues," *Conscience* 7, 1: 1–10.

Mastroianni, Luigi, Jr., et al. (eds.). 1990. *Developing New Contraceptives: Obstacles and Opportunities*. Washington, DC: National Academy Press.

*The New York Times*. 1988. 29 October.

*The New York Times*. 1989. 26 March; 12 August.

*The New York Times*. 1990. 9 March; 8 November.

Oakley, Deborah. 1975. Memorandum to the World Federation of Public Health Organizations and attached "Observations on population politics at the International Conference on Women, Mexico City, June 1975." Ann Arbor: University of Michigan, unpublished.

Paige, Connie. 1983. *The Right to Lifers*. New York: Summit Books.

Petchesky, Rosalind P. 1984. *Abortion and Woman's Choice: The State, Sexuality and Reproductive Freedom*. New York: Longman.

Pope Paul VI. 1968. *Humanae Vitae*. Boston: Daughters of St. Paul.

Requena, M. 1970. "Abortion in Latin America," in *Abortion in a Changing World*, ed. R. E. Hall. vol. I. New York: Columbia University Press, pp. 338–352.

Rosenfield, Allan, et al. (eds.). 1989. "Women's health in the Third World: The impact of unwanted pregnancy," *International Journal of Gynaecology and Obstetrics* Supplement 3:

Schattschneider, E. E. 1960. *The Semi-Sovereign People*. New York: Holt, Rinehart and Winston.

Symonds, Richard and Michael Carder. 1973. *The United Nations and the Population Question, 1945–1970*. New York: McGraw-Hill.

Trueman, Patrick. 1980. "A warning to Latin America," in *The Dignity of Man and Creative Love: Selected Papers from the Congress for the Family of the Americas, Guatemala, July 1980*, eds. Joseph Santamaria, Pedro Richards, and William Gibbons. New Haven: Knights of Columbus.

United Nations. 1975. "World Population Plan of Action," *Population and Development Review* 1, 1: 163–181.

———. 1984. "Recommendations for the further implementation of the World Population Plan of Action," *Population and Development Review* 10, 4: 758–782.

United Nations Fund for Population Activities (UNFPA), the International Planned Parenthood Federation, and the Population Council. 1981. *Family Planning in the 1980's: Challenges and Opportunities*. New York: UNFPA.

———. 1990. "Relevant facts about the United Nations Population Fund (UNFPA)." New York: UNFPA.

United States. 1984. "US policy statement for the International Conference on Population," *Population and Development Review* 10, 3: 574–579.

Warwick, Donald. 1980. "Foreign aid for abortion," in *Abortion Parley,* ed. James T. Burtchaell. Mission, KS: Andrews and McMeel.

———. 1982. *Bitter Pills: Population Policies and Their Implementation in Eight Developing Countries*. Cambridge: Cambridge University Press.

Women's Global Network for Reproductive Rights (WGNRR). 1985–92. *Newsletters*. Amsterdam.

World Health Organization (WHO). 1978. *Induced Abortion: Report of a WHO Scientific Group*. WHO Technical Reports Series No. 623. Geneva: WHO.

———. 1989. Newsletter on the WHO-HRP research initiative on the determinants and consequences of induced abortion. Geneva: WHO.

———. 1990. "WHO statement on RU486." Geneva: WHO Division of Public Information and Public Relations.

———. 1991. *From Abortion to Contraception: Report of the Tbilisi Expert Meeting*. Geneva: WHO.

Zatuchni, Gerald I., et al. (eds.). 1979. *Pregnancy Termination: Procedures, Safety, and New Developments*. Hagerstown, MD: Harper and Row.

# POSTSCRIPT

# The Politics of Family Planning: Issues for the Future

C. ALISON MCINTOSH
JASON L. FINKLE

THROUGHOUT THIS VOLUME we have used the term population policy as synonymous with fertility limitation. We are aware, of course, that demography embraces not only fertility but also mortality and migration. Each of these variables is an important matter for analysis, but each generates its own set of political problems and issues that overlap only to a limited degree. The politics of fertility is different from the politics of mortality and migration and, as has been demonstrated above, can be analyzed separately. Without intending to diminish the significance of the other demographic variables, we believe that fertility policy warrants the singular attention we have given it in this book. Moreover, fertility reduction remains one of the most intractable and contentious problems facing the global community. We expect this situation to continue in the future.

The central thesis of the volume is that while population policy continues to generate political differences, the nature of the debate in recent years has changed in a fundamental way. Starting in Asia in the 1950s, political leaders in developing countries gradually abandoned the historical assumption that a large and growing population was a necessary basis for economic development. In its place arose the radical notion that rapid population growth might constitute an impediment to economic development. Despite the beginnings of a fundamental change in the perceptions of the costs and benefits of population growth, most governments continued for some time to think in geopolitical terms when contemplating the adoption of population policies. The difference today is that while almost all leaders of developing countries recognize the advantages to be gained from reducing population growth, the primary means for achieving this result—family planning programs—has become a matter of heated political controversy. New actors have entered the arena and new issues have been raised. In short, the politics of population has entered a new phase as it has become more complex and more sophisticated.

While organized family planning programs have come to occupy center stage in the politics of population, to assume that the "old" population issues have vanished would be unwise. The controversies and conflicts that arose between the sixteenth and nineteenth centuries as the nation-state emerged, both as a concept and a reality, are reappearing, mainly at the subnational level. Thus, even in Great Britain, the country in which the process of nation building is generally considered to have proceeded the farthest, the post–World War II period has witnessed several outbursts of Scottish and Welsh nationalism, with demands being made for more autonomy from England. In the much younger nation of Canada, repeated calls by the Quebecois for the separation of their province from anglophone Canada, have on more than one occasion precipitated serious constitutional crises. In Belgium, a quarter-century of growing tensions between French-speaking Wallonia and Flemish-speaking Flanders has forced the country to accept a constitutional reform creating a federal structure.[1]

Just what forces are at work in these cases of reasserted nationalism is not entirely clear. What does seem evident is that subnational groups that feel themselves to have been neglected, discriminated against, or ignored in the political or economic affairs of the nation have been moved to reaffirm their ancient cultural identities and to claim equal or separate status. Somewhat similar forces appear to underlie the resurgence of religious fundamentalism in the late twentieth century. Not only has fundamentalism become a powerful and potentially destabilizing force among Muslims, but it has also appeared in almost all religions ranging from Christianity to Hinduism. In all these cases there seems to be a sense that the forces of secularism and materialism are out of control and that stability can be regained only by a return to "traditional" religious values. To varying degrees all these fundamentalist religious movements have a political dimension.

At the present time, large areas of the world hold the potential for the emergence of nationalist, religious, or communal conflict in which demographic factors may come to play a significant role. The most dramatic precipitant of subnational conflict has been the recent and sudden collapse of communism in the former Soviet empire. In this vast area, stretching from central and eastern Europe to the shores of the north Pacific Ocean, countless national groups were denied the expression of their identities for some 70 years. As we have seen in the former Yugoslavia, the sudden lifting of repression has unleashed the nationalistic and religious turbulence that has smoldered for centuries in the region.[2] Sporadic but violent conflict has also appeared in Armenia and Georgia, and among some of the newly independent states of central Asia.

The situation in those areas of the world that were subject to colonial domination until the end of World War II has many parallels with that of the former Soviet empire, but the forces at work have been more subtle and less visible. In many ex-colonial countries, notably in the former French Indochina, on the Indian subcontinent, and in Africa, conflict among subnational groups was kept in check by the colonial presence. Once the colonial powers departed, conflicts have arisen from time to time, especially in those countries that have not experienced the rapid economic growth characteristic of the newly industrialized states of southeast Asia.

As we have witnessed in recent years, the countries of sub-Saharan Africa stand out as fertile ground for the emergence of ethnic and tribal conflict. The sense of nationalism that united the tribes within colonial countries in their struggle against the metropolitan powers did not long outlive the achievement of independence. In the absence of a unifying force, most African leaders ruled through traditional patrimonial systems that did little to erode tribal and ethnic solidarity (e.g., Jackson and Rosberg, 1982). There is evidence, moreover, that the emerging participatory polities that began to appear in the towns and cities during the decade prior to independence quickly faded as many political leaders turned to one-party government and personal rule (Kafsir, 1976). In short, in many sub-Saharan African countries, 30 years of independence have done little to forge a sense of nationhood or to develop alternatives to tribalism as the basis of political participation—a situation that will only be overcome through the exercise of great political skill.

Many observers have noted that rather than heralding an era of peace and prosperity, the end of the cold war has ushered in a period of political, economic and social unrest. Much of this turmoil seems to derive from the shattering of relationships among and within nations that seemed dangerous at the time but that now appear to have been relatively stable and predictable.

Changes of this magnitude cannot take place without leaving their mark on demographic trends and the ways in which they are perceived by political elites. Two types of consequences seem especially likely. First, as we have suggested above, the reaffirmation of identity by ethnic, linguistic, religious, or communal groups within existing countries seems designed to foster the emergence of demographic-political rivalries. In places where the consensus on the need to reduce the rate of population growth is of recent origin or frail, a return to pronatalist attitudes among subnational groups could weaken commitment to fertility regulation and reduce resources devoted to family planning.

A second consequence is the enormous increase in the number of refugees and other emigrants fleeing areas engulfed by civil strife. In Europe especially, the level of immigration has already reached the limits of tolerance in recipient countries, a number of which are tightening their entry criteria (*New York Times*, 1993). In recent years, international law has tended increasingly to be interpreted as supporting the right of individuals freely to leave their own countries while limiting the freedom of states to control the influx of migrants.[3] Clearly, international migration will become an even more serious problem if civil strife within nations persists.

## The ethics of family planning programs

Ironically, as the last cluster of developing countries is becoming convinced of the need to lower fertility and to widen the acceptance of contraception, a number of questions have been raised concerning the legitimacy of family planning programs as they are currently structured. For the most part, these questions have been advanced by individuals and organizations in the more developed countries, although, to some extent, they are also being raised in developing countries. In recent years, almost every aspect of family planning programs has been subjected to critical scrutiny from an ethical perspective. This examination, far exceeding that directed to any other development program, has considered issues ranging from the types and modes of action of contraceptives, to the ways in which family planning is promoted and services are delivered, the quality of services offered, and whether fertility reduction is an appropriate objective. Always present in this debate are the thorny issues of induced abortion and sterilization and of whether contraceptive services should be offered to unmarried adolescents.

Family planning programs have always been the subject of political-ethical debate. The inherently sexual nature of contraception and the private nature of reproductive decisionmaking are sufficient, of themselves, to ensure that government family planning services will always arouse political sensitivities. In recent years, however, the intensity and scope of the ethical critique has vastly exceeded that of earlier debates, even though fertility regulation and family planning programs are more widely accepted than ever before in developing countries. The significant question is why this change has occurred.

The recent escalation of the ethical critique of family planning programs has its origin in two distinct but intersecting movements—the New Right coalition of the Reagan and Bush administrations and, more broadly, the international feminist movement. The New Right coalition brought together conservative Republicans, elements of the Catholic Church, Protestant fundamentalists, and other "right-to-life" advocates, in a concerted attack on family planning and abortion both in the United States and elsewhere (Crane and Finkle, 1987). Underlying all the specific targets of the New Right campaign—abortion, sterilization, services for adolescents, allegedly coercive practices in US-supported family planning programs overseas—lay a foundation of moral and religious values subscribed to by some of the most conservative members of US society. As is well known, the coalition found a willing ear in the Reagan White House that permitted the US debate to continue, made no effort to halt the growing violence against abortion providers in the United States, and took a number of actions that undermined US international population assistance.[4]

Compared with the New Right, the international feminist movement has much deeper historical roots and more complex objectives.[5] Historically, feminists in the United States have been strong supporters of international family planning programs. Along with most family planning providers, feminists have regarded women's ability to choose the timing and spacing of their children as one of the most fundamental means of enabling them to assume broader social and economic roles, and freeing them from the hazards of frequent childbearing (e.g., Germain, 1975). Unlike members of the New Right, feminists today seek to ensure that women everywhere have access to a full range of contraceptive services, including legal abortion and sterilization. But while feminists firmly oppose the conservative agenda and may have drawn strength from opposing it, in pursuit of broader objectives they have mounted their own ethical critique of family planning programs.

The internationalization of the feminist movement was greatly facilitated by the UN Decade of Women, 1975–85, and the emergence of such international organizations as the International Women's Health Coalition based in New York, the Women's Global Network on Reproductive Rights in Amsterdam, and ISIS-International, in Geneva and Rome. Through these and similar organizations, feminists who support family planning as a goal in itself have made common cause with women whose chief aims are to halt the oppression of women in all its forms including the neglect of women's health, to establish women's rights, and to promote their status and dignity. While a majority of feminists still regard the promotion of access to contraceptive supplies and counseling as a worthwhile objective, increasing numbers are questioning the delivery of family planning services that are independent of comprehensive health services and that are provided without medical supervision. A growing radical wing disapproves the use of long-term, surgical, or hard-to-reverse methods on the grounds that they diminish women's control of their reproductive lives. A tendency exists in some women's organizations to adopt absolutist standards to measure the impact of hormonal methods on women's health. This stance is opposed to the "scientific" view that the correct standard is the relative risk as compared with the risk of pregnancy and childbearing in many less developed countries. The use of incentives to encourage adoption of family planning, long a subject of ethical debate, is being questioned again in both the conservative and the feminist camps. One of the most serious feminist criticisms is that programs with demographic targets subject women's needs to impersonal, and therefore inappropriate, societal goals.[6]

The broad range of issues receiving scrutiny from an ethical perspective is thus addressed by individuals and groups that span the political spectrum. Conflicting pressures are being brought to bear on family planning programs to bring them into conformity with religious, moral, human rights, health, and (more broadly) feminist values. While these issues are being raised at policymaking levels within the international population community, such value questions cannot be settled in the abstract. Rather, they will be played out in a myriad of local and central programmatic decisions affecting implementation as well as policy. Moreover, as politics is an unending process in which values carry an extra punch, we anticipate that ethical questions will continue to shape the politics of family planning for some time to come.

## And what of population?

Fairly wide agreement exists that, overall, population assistance has been one of the most effective areas of development aid. Not only has international population assistance succeeded in increasing access to contraceptives and supporting services, it has also contributed to the decline in birth rates documented in much of the developing world (Cassen and Associates, 1986; Ness and Ando, 1984). Despite the substantial decline of overall fertility in developing countries, however, population growth will continue to be a problem for the future. Although world population growth has been reduced from a high of 2.1 percent per annum in 1965–70, to 1.7 percent in 1985–90, the United Nations medium projection shows that 15 years from now, the global population will still be growing by more than 90 million people each year (United Nations, 1992). Most of this growth will be in developing nations. If population stabilization remains an objective of the international community, a strong need exists to commit substantial—and increasing—funds to the reduction of fertility levels. If funds and commitment can be increased, a successful outcome is a reasonable probability.

Since the early 1960s, the rationale for organized family planning programs has undergone a number of changes, reflecting the rise and decline of competing theories of fertility and assessments of the consequences of rapid population growth. As rationales have come and gone, their influence has been felt in modifications of policy and programs and also in levels of donor commitment and funding. To a considerable extent, family planning programs have evolved in ways designed to anticipate and address many of the issues now being raised by the critics of family planning. A brief review of this history might help to place the ethical critique in perspective.

In the first chapter of this volume we noted that in the 1950s and 1960s, various arguments were advanced to justify intervention to lower the rate of population growth. The most persuasive of these were clearly economic.[7] Faced with an obvious gap between expectations and reality in the rate of socioeconomic development in the less developed countries, the richer countries felt a need to intervene, not only to lessen the drag of population growth on development, but also to protect their own investments in development aid. The richer countries were supported in their decision by some Asian governments that also identified population growth as a major impediment to their development. While there has never been unanimity among economists on the relationships between population and economic growth, the consensus among them that rapid population growth is a handicap to development was probably at its peak in those decades.

Inspired by this perceived economic imperative, the newly established family planning programs searched for ways to overcome the tremendous resistance to lowering fertility. In their attempt to promote family planning practice while handicapped by inadequate technologies, the new programs undoubtedly violated many ethical principles that have since been recognized. Usually they offered a minimal choice of methods. Oral contraceptives contained far higher levels of hormones than do today's preparations, and they were proffered with less attention to their potential side effects on women's health than would be acceptable today. Many programs were target-oriented, and numerous experiments were undertaken with incentives and disincentives that posed serious ethical questions.[8] Since this period, however, family planning programs have largely abandoned the strictly categorical form that was reminiscent of a military campaign against communicable disease. A wider range of safer contraceptive methods is now available, and new methods of service delivery have been devised that bring contraceptive supplies closer to women's homes. Greater attention is being given to educating and counseling women, thus improving their ability to make real choices, and more attention is given to the treatment of side effects. Increasingly, family planning services are integrated with a range of related health services, and often with income-generating and training programs for women. While much remains to criticize in many programs, there is no doubt that the quality of services has greatly improved. Indeed, improving the quality of care has recently become the subject of a concerted effort by family planning providers worldwide (Bruce, 1990; Jain, 1992; Mensch, 1993).

Most of these changes have taken place in the last decade, during which the international population movement has regained strength after a period in which it seemed to lose a sense of direction. The precipitating cause of this malaise was the attack mounted by a coalition of third world countries, led by Algeria and Argentina, at the World Population Conference held at Bucharest in 1974. At this meeting, third world countries joined together to argue forcefully that fertility reduction in the developing world was an imperialist plot to avoid the necessity of investing in development, and to preserve the structure of economic relations between North and South.[9] While this was clearly the most critical factor undermining the confidence of the population establishment, it was buttressed by such other forces as a growing recognition that the widespread distribution of contraceptives was insufficient by itself to motivate women to accept them. The views of economists who were questioning the underlying assumption that population growth is inimical to economic development were also important (e.g., Simon, 1977 and 1981). Several years were required for the population establishment to regroup after these attacks on its basic tenets.

Paradoxically, it was under the Reagan administration, arguably the most hostile toward population control of any US presidency, that international family planning programs started to regain their vigor. Despite the formidable attack of the New Right coalition, and the financial blows to the United Nations Population Fund and the International Planned Parenthood Federation, the US family planning enterprise was enabled to reenergize its own operations and to infuse funds and technical assistance into indigenous family planning planning agencies in developing countries. The key to this transformation, albeit inadvertent, was undoubtedly the Reagan administration's identification of the private sector as the proper locus of social and economic growth. In the family planning field, this new focus has fostered the multiplication of voluntary and for-profit agencies in developing countries, has unleashed the enthusiasm and energies of innumerable talented women and men, and directed them to the development of local family planning programs. A productive partnership between private agencies in donor and recipient countries is providing a mechanism through which family planning, health, education and training programs, and income-generating schemes can be integrated at the service level while retaining sophisticated and specialized administration at higher levels. Such programs in the private sector may well provide a model for government services in countries where these are the preferred mode of operation. While these innovative programs are still in their infancy, they have provided a much-needed stimulus and have the potential to revolutionize the delivery of contraceptive services on a wider scale than previously was possible.[10]

In September 1994, the United Nations will convene the third decennial intergovernmental conference on global population problems. The theme of the conference, which is to be held in Cairo, is reflected in the official name of the meeting: the International Conference on Population and Development. As was the case in the two previous conferences, formal preparations include six Expert Group Meetings, intergovernmental meetings hosted by the UN Regional Economic Commissions, and three meetings of the special Preparatory Committee.

To the political analyst, one of the more engaging features of these intergovernmental conferences is their propensity to assume a life of their own and to move the discussions in unanticipated directions. While the meetings are inspired by a desire to take stock, resolve differences, and reinforce the sense of common purpose, the highly political nature of population issues endows them with a potential for redefinition by the participants during the course of the conference. At both Bucharest and Mexico City, the formal objectives of the conference were achieved to a considerable extent. At the same time, the meetings raised new issues and suggested new ways of approaching them. Once the formalities were over, these new ideas and issues set the agenda for the next decade.

The issues for the Cairo meeting are many and varied. Some of them, including abortion and sterilization, the ethical critique, the feminist agenda, and demographic objectives, have been touched on in this volume. Immigration, human rights, and the environment are also likely to receive attention. How these issues will be treated by the participants at the conference, what connections will be made among them, remains to be seen. One thing is clear, however. At the conference, and in intergovernmental circles in general, all of these will be issues of "high" politics; by contrast, their implications will be played out in the everyday "low" politics of family planning implementation.

## Notes

1 A series of constitutional crises coming at short intervals during the 1980s was only resolved by the approval in 1988 of a plan to create a federal structure. The third of three stages was completed in June 1993. Belgium is now divided into three linguistic/communal entities: French-speaking Wallonia, Flemish-speaking Flanders, and bilingual Brussels. See *Keesing's Record of World Events* (1989 and 1991) and *The Economist* (1993). A major cause of the crises was a change in the relative prosperity of Wallonia and Flanders when Wallonia's economy, based on traditional industry, declined in relation to that of the more modern economy of Flanders.

2 See Besemeres, 1980, Part IV, pp. 157–246, for detailed and insightful analysis of the ethnodemographic conflict in Yugoslavia.

3 The noted legal scholar Richard Plender (1988: 119), reviewing the status of migration law, states that "the right to leave any country, including one's own, is guaranteed expressly in a series of multilateral treaties ratified by the greater part of the international community." Important relevant documents include the International Covenant on Civil and Political Rights, which by 1985 had been ratified by 83 states; the Convention on the Elimination of All Forms of Racial Discrimination (ratified in 1985 by 124 states); and the constitutions of numerous countries. Similarly, modern legal scholars argue that the principle "that States are free to control at will the entry and residence of aliens," is no longer tenable in light of contemporary international interdependence. See for example Plender, 1988, pp. 1–4 and Goodwin-Gill, 1978, p. 57.

4 Since the Helms amendment of 1973, no US funds have been used to promote or carry out abortions overseas. In what became known as the Mexico City Policy, the Reagan administration prohibited the transfer of US funds to any agency, US or foreign, that provided abortions even with non-US government monies. Also, after the 1984 Mexico City Conference, the Reagan administration withheld funds from the International Planned Parenthood Federation because some of its affiliates in third world countries provided abortions. The Reagan administration also withheld its annual donations to the

United Nations Population Fund because of the Fund's presence in China, which allegedly supported coercive abortion policies. The Clinton administration is taking steps to restore the funding to both UNFPA and IPPF.

5 In the United States, a radical feminist movement protested Margaret Sanger's decision to make birth control more respectable by inviting the participation of physicians in her clinics, and by changing the name from birth control to family planning. Radical feminists have emerged from time to time to protest the loss of women's control over contraceptive services and to press for a more holistic approach to women's reproductive health (e.g., Gordon, 1990 and Reed, 1978).

6 Some indication of the range of feminist interests and activities can be gained by a review of the joint policy statements issued by feminist organizations and mainline multilateral and bilateral agencies. These include: Germain and Ordway (1989) and World Health Organization and International Women's Health Coalition (1991). Other statements include Petchesky and Weiner (1990) and Berer (1991). Many feminist aims have been promulgated in the form of a Women's Declaration on Population Policies, prepared for the International Conference on Population and Development to be held in Cairo in September 1994. The Declaration was prepared by 25 "initiators" representing women's organizations from all parts of the world, coordinated by the International Women's Health Coalition.

7 This point is amply made by Symonds and Carder (1973) for the international system as a whole. The US National Security Council Memorandum (NSSM 200), which in December 1974 set out the US justification for reducing population growth, touches on resource consumption and depletion, health and welfare issues, and potential political consequences, but the economic concerns are clearly foremost. The Executive Summary of this memorandum is reprinted in *Population and Development Review* 8, 2: 423–433.

8 Donald Bogue's 1967 article predicting that family planning programs will have brought the population crisis to an end by the year 2000 is an example of the faith in a technological solution that was prevalent at the time. USAID's "inundation" approach to contraceptive delivery, based on what was thought to be an almost universal demand for family planning, is a good example of the aggressive, centralized, single-purpose program favored by some at that time (see Ravenholt and Chao, 1974).

9 For an analysis of the conference and the positions taken by the contending parties see Finkle and Crane, 1975. The draft World Population Plan of Action prepared in advance by the United Nations and a group of Western and Asian countries was substantially modified—and, from a strictly demographic perspective, weakened—during the conference. The draft plan's stipulation for developing countries to establish quantitative targets for fertility reduction was removed. The concepts of demographic and contraceptive acceptor targets are open to abuse at the service-delivery level. However, some quantitative target is necessary, if only to provide a basis for calculating the number of contraceptives that will be needed.

10 Some population assistance has always gone to voluntary agencies in recipient countries. In the early days, when recipient country governments were reluctant to become involved in family planning activities, voluntary agencies received most of the available funding. During the past decade, however, the amount of work done through indigenous private agencies, both voluntary and for-profit, has greatly increased.

## References

Berer, Marge. 1991. "What would a feminist policy be like?" *Conscience* 12, 5: 1–5.

Besemeres, John F. 1980. *Socialist Population Politics*. White Plains, NY: M.E. Sharpe.

Bogue, Donald J. 1967. "The end of the population explosion," *Public Interest* 7 (Spring): 11–20.

Bruce, Judith. 1990. "Fundamental elements of the quality of care: A simple framework," *Studies in Family Planning* 21, 2: 61–91.

Cassen, Robert, and Associates. 1986. *Does Aid Work? Report to an Intergovernmental Task Force*. Oxford: Clarendon Press.

Crane, Barbara B., and Jason L. Finkle. 1987 "The conservative transformation of population policy." *Governance, Harvard Journal of Public Policy* (Winter/Spring): 9–14.

*The Economist*. 1993. "The King is dead: Long live Belgium," 7 August, p. 47.

Finkle, Jason L., and Barbara B. Crane. 1975. "The politics of Bucharest: Population, development, and the new international economic order," *Population and Development Review* 1, 1: 87–114.

Germain, Adrienne. 1975. "Status and roles of women as factors in fertility behavior," *Studies in Family Planning* 6, 7: 192–200.

———, and Jane Ordway. 1989. *Population Control and Women's Health*. New York: International Women's Health Coalition, in cooperation with the Overseas Development Council.

Goodwon-Gill, Guy. 1978. *International Law and the Movement of Persons Between States*. Oxford: Clarendon Press.

Gordon, Linda. 1990. *Woman's Body, Woman's Right: Birth Control in America*. Revised edition. New York: Penguin Books.

Jackson, Robert H., and Carl G. Rosberg. 1982. *Personal Rule in Black Africa*. Berkeley, Los Angeles, and London: University of California Press.

Jain, Anrudh K. (ed.). 1992. *Managing Quality of Care in Population Programs*. West Hartford: Kumarian Press.

Kafsir, Nelson. 1976. *The Shrinking Political Arena: Participation and Ethnicity in African Politics with a Case Study of Uganda*. Berkeley and Los Angeles: University of California Press.

*Keesing's Record of World Events*, 1989, 35, 6: 36,774; 1991, 37, 2: 38,465.

Mensch, Barbara S. 1993. "Quality of care: A neglected dimension," in *The Health of Women: A Global Perspective*, eds. Marjorie Koblinsky, Judith Timyan, and Jill Gay. Boulder, CO: Westview Press, pp. 235–253.

Ness, Gayl D., and Hirofumi Ando. 1984. *The Land Is Shrinking: Population Planning in Asia*. Baltimore and London: Johns Hopkins University Press.

*The New York Times*. 1993. "Western Europe is ending its welcome to immigrants," 10 August, p. 1.

Petchesky, Rosalind P., and Jennifer A. Weiner. 1990. *Global Feminist Perspectives on Reproductive Rights and Reproductive Health: A Report on the Special Sessions held at the Fourth International Interdisciplinary Congress on Women, Hunter College, New York City*. 3–7 June, mimeo.

Plender, Richard. 1988. *International Migration Law*. Dordrecht, Boston, and London: Martinus Nijhoff.

Ravenholt, R. T., and John Chao. 1974. "Availability of family planning services is the key to rapid fertility reduction," *Family Planning Perspectives* 6, 4: 217–223.

Reed, James. 1978. *From Private Vice to Public Virtue*. New York: Basic Books.

Simon, Julian. 1977. *The Economics of Population Growth*. Princeton, NJ: Princeton University Press.

———. 1981. *The Ultimate Resource*. Princeton, NJ: Princeton University Press.

Symonds, Richard, and Michael Carder. 1973. *The United Nations and the Population Question, 1945–1970*. New York: McGraw-Hill.

United Nations. 1992. *World Population Monitoring Report 1991*. Population Studies No. 126. New York: United Nations.

United States National Security Council. 1974. Memorandum (NSSM 200). Executive summary, reprinted in *Population and Development Review* 8, 2: 423–434.

*Women's Declaration on Population Policies*. 1993. New York: International Women's Health Coalition.

World Health Organization, Special Programme of Research, Development and Research Training in Human Reproduction, and the International Women's Health Coalition. 1991. *Women's Perspectives on the Selection and Introduction of Fertility Regulation Technologies*. Geneva: WHO.

# AUTHORS

CARLOS ARAMBURÚ is Regional Vice President for Latin America, Pathfinder International, Regional Office for Latin America.

GUSTAVO CABRERA is Director, Center for Demographic and Urban Development Studies, El Colegio de México, Mexico City.

JOSEPH CHAMIE is Director, Population Division, Department for Economic and Social Information and Policy Analysis, United Nations, New York.

BARBARA B. CRANE is International Population Fellow, University of Michigan.

RUTH DIXON-MUELLER is Consultant, International Women's Health Coalition, New York.

JASON L. FINKLE is Professor of Population Planning, University of Michigan.

ADRIENNE GERMAIN is Vice President, International Women's Health Coalition, New York.

MERILEE S. GRINDLE is Fellow of the Institute, Harvard Institute for International Development.

CHARLES B. KEELY is Donald G. Herzberg Professor of International Migration and Professor of Demography, Georgetown University, Washington, D.C.

OMARI H. KOKOLE is Associate Director, Institute of Global Cultural Studies, State University of New York at Binghamton.

ALI A. MAZRUI is Director, Institute of Global Cultural Studies, and Albert Schweitzer Professor in the Humanities, State University of New York at Binghamton.

C. ALISON MCINTOSH is Director, Population Fellows Program, University of Michigan, and teaches in the Department of Population Planning and International Health.

V. A. PAI PANANDIKER is Director, Center for Policy Research, New Delhi.

JOHN W. THOMAS is Institute Fellow, Harvard Institute for International Development.

P. K. UMASHANKAR is former Special Secretary, Department of Family Welfare, Government of India, New Delhi.

DONALD P. WARWICK is Institute Fellow, Harvard Institute for International Development.

TYRENE WHITE is Associate Professor of Political Science, Swarthmore College.